T0189341

Next Generation Multilayer Graded Bandgap Solar Cells

A. A. Ojo • W. M. Cranton • I. M. Dharmadasa

Next Generation Multilayer Graded Bandgap Solar Cells

 Springer

A. A. Ojo
Sheffield Hallam University
Sheffield, UK

I. M. Dharmadasa
Sheffield Hallam University
Sheffield, UK

W. M. Cranton
Sheffield Hallam University
Sheffield, UK

ISBN 978-3-030-07230-8   ISBN 978-3-319-96667-0   (eBook)
https://doi.org/10.1007/978-3-319-96667-0

This Springer imprint is published by the registered company Springer Nature Switzerland AG.
The registered company address is: Gewerbestrasse 11, 6330 Cham, Switzerland

Preface

Direct conversion of light energy into electrical energy or photovoltaic technology has continually developed over the past five decades. Solar panels based on silicon and thin-film solar panels based on CdTe and CuInGaSe$_2$ are now in the market. The cost of solar panels have reached ~1.0 \$W^{-1}, and further reduction to ~0.5 \$W^{-1} will enable this technology to become a main stream energy supply in the future. Scientific research in this field should therefore be directed towards next-generation solar cells. Key features of these solar cells should be low cost of manufacturing, high conversion efficiency and durability over a period of three decades. Availability of materials required and their non-toxic nature are also important factors.

High conversion efficiencies can only be achieved by harvesting photons from all energy ranges, across the ultraviolet, visible and infrared radiation regions. Devices with many bandgaps have been proposed in the early 1960s, but experimental attempts were scarce. There are few reports in the literature on grading of one layer of a *p-n* junction and achieving improved device parameters. However, work has not progressed forward in order to develop high performing devices. One of the authors of this book (IMD) published graded bandgap devices based on II–VI materials in 2002 and improved this idea to fully graded devices between the front and back electrical contacts in 2005. These devices were experimentally tested during the same year to achieve outstanding device parameters confirming the validity of the new designs. These fully graded devices also benefit from "impurity PV effect" and "impact ionisation" to enhance photo-generated charge carriers. With the experimental confirmation, authors focussed their work on graded bandgap devices based on low-cost, scalable and manufacturable electroplated materials.

This book covers several important areas in the field. The book summarises the results of electroplating of semiconductors and details on three main solar energy materials: ZnS as a buffer layer, CdS as the window layer and CdTe as the main light-absorbing material. Growth details and material characterisation using most appropriate techniques are presented. This will serve as a handbook for new and established researchers to continue work in their research fields.

This book will also serve as practical reference for graded bandgap device fabrication and assessment. The work presented in this book shows the achievement of 15–18% conversion efficiencies for lab-scale devices utilising electroplated materials. Authors believe that systematic work along this line could produce efficiencies close to mid-20%. The knowledge gained from this work can also be equally applied to other thin-film solar cells based on $CuInGaSe_2$, kesterite and perovskite materials.

Electrodeposition is a low-cost but very powerful technique as a semiconductor growth technique. Continuation of this exploration will lead to develop large-area electronics (LAE) sector in the future. In addition to large-area solar panels, electrodeposition will enable to develop large-area display devices and numerous other devices based on nanotechnology.

Sheffield, UK   A. A. Ojo
May 2018   W. M. Cranton
                                                                                  I. M. Dharmadasa

Acknowledgement

The achievements made in this work would not have been possible without the grace and blessings of God who makes all things beautiful in His time. Tremendous appreciation goes to my director of studies (DOS), Prof. I.M. Dharmadasa, for his professional mentorship. I do also recognise my second supervisor, Prof. Wayne Cranton, Dr. A.K. Hassan and Dr. Paul Bingham for their contributions.

Sincere appreciation goes to all the members of the Solar Energy Research Group of Sheffield Hallam University; this includes Dr. O.K. Echendu, Dr. F. Fauzi, Dr. N.A. Abdul Manaf, Dr. H.I. Salim, Dr. O.I. Olusola, Dr. M.L. Madugu, Dr. Burak Kadem and Dr. Yaqub Rahaq for their useful advice, technical discussions and constructive criticisms. Appreciation also goes to the members of staff at MERI including Gillian Hill, Jayne Right, Gail Hallewell, Rachael Toogood, Clare Roberts, Corrie Houton, Gary Robinson, Deeba Zahoor, Stuart Creasy, Paul Allender, Bob Burton and Anthony Bell for their administrative and technical support during my research program. I do also acknowledge the contributions of my family and friends within and outside MERI most especially Moyo Ayotunde-Ojo, the Kehinde Ojo's and the Ajiboye's. The support of the VC, DVCs, Dean of Engineering, HOD Mechanical Engineering and other departmental and faculty colleagues in Ekiti State University (EKSU), Ado-Ekiti, Nigeria, is also recognised.

Ayotunde Adigun Ojo

Contents

Chapter 1
Introduction to Photovoltaics

1.1 Global Energy Supply and Consumption

Energy is an essential constituent of economic growth and development, the demand for which increases with a corresponding increase in population [1]. With global population growing from 1 billion in the 1600s to 7.5 billion at present (2017) and a projected increase to 9.7 billion by 2050 [2], concerns over exhaustion, energy resources supply difficulties and substantial environmental impacts (such as depletion of the ozone layer, global warming, climate change, amongst others) have been raised for conventional energy sources [3, 4]. The 2017 edition of British Petroleum's annual outlook shows that fossil fuel has dominated the world's energy resource accounting for ~85% of the total consumption, with ~5% from nuclear power and less than 10% from renewable energy resources [5]. This trend cannot be sustained without any catastrophic effect [3, 4] with increasing energy demand. Hence, there is a global imperative to move towards carbon-neutral energy source solutions commensurate with or greater than the present-day energy demand.

1.2 Energy Sources

Depending on how long it takes for a primary energy source to be replenished, energy sources can be categorised as either non-renewable or renewable (or alternative) energy sources.

1.2.1 Non-renewable Energy Sources

Non-renewable energy sources are sources that are not replenishable within a human lifetime. Energy sources such as fossil fuels (coal, crude oil, natural gas and other

© Springer International Publishing AG, part of Springer Nature 2019
A. A. Ojo et al., *Next Generation Multilayer Graded Bandgap Solar Cells*,
https://doi.org/10.1007/978-3-319-96667-0_1

petroleum products) and uranium (for nuclear energy) fall under the category of non-renewable energy sources [6]. Fossil fuels are generated from combustible geological deposits of organic materials formed from decayed animals and plants buried in the earth's crust and subjected to high pressure and temperature over thousands and millions of years. The combustion of fossil fuel products or by-products releases a tremendous amount of greenhouse gases. Uranium ore which serves as fuel in nuclear power plants is classified as non-renewable due to its rarity, as it is mined in limited locations around the globe.

1.2.2 Renewable Energy Sources

Renewable or alternative energy sources are sources that are replenishable within a human lifetime. Sunlight, wind, hydro, ocean waves and tide, geothermal heat and biomass are the primary energy sources which fall under this classification due to their continuous availability and reusability. It is interesting to note that most of the renewable energy resources other than geothermal energy and tidal wave energy depend directly or indirectly on sunlight. With about 4.3×10^{20} J [7] and an estimated 1367 Wm^{-2} irradiance made available to the earth's surface through sunlight, solar energy emerges by far as the most abundant exploitable resource. Based on these facts, significant research and development effort has been deployed into the science and technology of achieving high solar energy conversion efficiency and reduction in the cost of production.

1.3 Solar Energy

With a mass of about 2×10^{30} kg, a diameter of 1.39×10^9 m, a surface temperature of about ~6000 K and a core temperature of about ~1.5×10^7 K, the sun stands as the primary source of solar energy and the centre of the solar system [8]. The energy generated by the sun is achieved by the constant fusion of hydrogen to helium nuclei and the release of a significant amount of energy (in the form of electromagnetic radiation and heat) in a process known as the thermonuclear fusion. The radiation from the sun reaches the earth at ~500 s putting into consideration the speed of light $(2.99 \times 10^8 \, ms^{-1})$ and the mean distance between the earth and the sun $(1.496 \times 10^8$ km). The average amount of radiation measured at the sun's surface is about $5.691 \times 10^7 \, Wm^{-2}$, with the irradiance reaching the earth's atmosphere being ~1367 Wm^{-2} [9]. Fifty-one percent of the incident radiation reaches the earth's surface with enough energy in 1 h to cater for global energy utility in a year [7, 10].

The remaining radiation is accounted for by the reflection of the incident radiation back into space and absorption by the atmosphere with each of these factors valued at about 30% and 19%, respectively [11], as shown in Fig. 1.1. The high attenuation of the solar radiation as it reaches the earth as compared to the radiation on the surface of the sun can be attributed to the effect of air mass (AM).

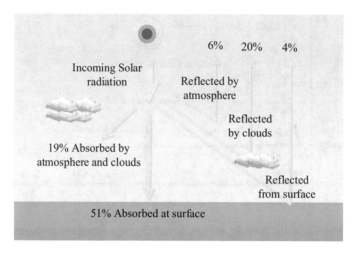

Fig. 1.1 Global modification of incoming solar radiation by atmospheric and surface processes

1.4 Air Mass Coefficients

Air mass (AM) is a measure of how sunlight propagates through the earth's atmosphere. It can also be defined as the shortest path through the atmosphere that sun rays pass through before reaching the surface of the earth. Air mass accounts for the attenuation of the radiation measured at the sun's surface compared to that measured at the earth's surface. The attenuation is due to absorption, reflection and scattering caused by the ozone layer (O_3), water molecules, carbon dioxide (CO_2), dust and clouds as sunlight pass through the atmosphere [9, 12]. Furthermore, the density of the atmosphere and the path length of the sunlight impact the attenuation of the radiation. Alternatively, air mass coefficient defines the optical path length through the earth's atmosphere, relative to the shortest path length vertically upwards, at the zenith. Air mass zero (AM0) refers to the standard spectrum outside the earth's atmosphere or the solar irradiance in space. The power density of AM0 is valued at 1367 Wm^{-2} [9]. This value is used for the characterisation of solar cells used in outer space. AM1.0 is used for tropical regions on earth surface where the sun is directly above the earth's zenith point. The incident power per area is valued at 1040 Wm^{-2} [9]. AM1.5 valued at 1000 Wm^{-2} or 100 $mWcm^{-2}$ [9] defines the power density of the incident solar radiation reaching the earth's surface known as insolation. This value is used by the PV industry as standard test condition (STC) for terrestrial solar panel characterisations.

1.5 Energy Distribution of the Solar Spectrum

As the solar radiation emanating from the surface of the sun reaches the earth's surface, its intensity and spectral configuration change due to attenuation. The spectral configuration known as the solar spectrum reaching the earth's surface

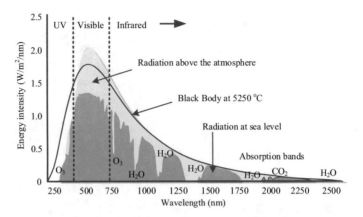

Fig. 1.2 The solar spectrum showing the spectral irradiance as a function of photon wavelength at the outer earth's atmosphere (black body), at the top of the atmosphere (AM0) and at the sea level (AM1.5) (Adapted from Ref. [13])

spans across the wavelengths (λ) of three spectra regions, namely, ultraviolet (UV), visible (Vis) and infrared (IR) as shown in Fig. 1.2 [13]. The ultraviolet region is approximately 5% of the total irradiation with a wavelength <400 nm, the visible region lies within the wavelength range of 380 and 750 nm, and it is approximately 43% of the irradiance. While the infrared region has a wavelength >750 nm, and it is about 52% of the irradiance distribution.

Based on requirements, solar energy technology has grown to focus on different spectral regions. Solar thermal technology is inclined to harness energy from the infrared region in the form of heat, while solar photovoltaic (PV) and concentrated solar (CS) power harnesses energy from both the visible and the ultraviolet spectral regions. However, recent development in PV has shown the possibility of harnessing energy from the UV, Vis and IR regions [14, 15]. In the research work presented in this book, emphasis will be laid on photovoltaic solar energy conversion technology using II–VI semiconductor materials.

1.6   Photovoltaic Solar Energy Conversion

Photovoltaic (PV) energy conversion technology is concerned with the direct conversion of solar energy (electromagnetic radiation from the sun) into electricity. The technology entails the generation of electrical power by converting solar radiation into a flow of electrons in the form of direct current (DC). Photons from solar radiation excite the electrons in a photovoltaic device into a higher state of energy, allowing them to act as charge carriers. The technology requires the use of suitable semiconductor materials with photovoltaic properties, the formation of a depletion region from which electron-hole (e-h) pairs are created provided photons with

Fig. 1.3 Schematic diagram of a simple *p-n* junction showing photon absorption and the effect of e-h collection through the external circuit

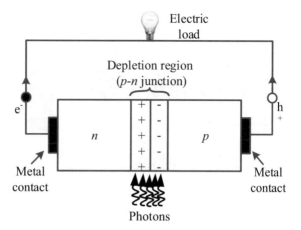

energy higher than the bandgap are introduced, efficient separation of oppositely charged carriers before recombination and transportation of the charge carrier through an external circuit [16]. For an excellent PV conversion, it is imperative that all these requirements are met. Figure 1.3 shows a schematic representation of the PV effect using a simple *p-n* junction configuration.

The absorber materials utilised in PV cell fabrication are categorised as first-, second- and third-generation (next-generation) solar cells. The first generation include monocrystalline silicon (mono-Si)- and polycrystalline silicon (poly-Si)-based solar cells [17]. These first-generation solar cells are the most established of all the solar cell categories. They are known for high material usage (bulk materials) and high fabrication cost. The second-generation solar cells incorporate thin-film technology with reduced material usage and material/fabrication cost, and they are scalable. Examples of second-generation solar cells include amorphous silicon (a-Si), cadmium telluride (CdTe) [18] and copper indium gallium diselenide (CIGS) [16, 19] thin-film solar cells. The third-generation solar cells are characterised by thinner films, low fabrication temperatures, high efficiencies and lower cost. They tend to overcome the Shockley-Queisser limit of power efficiency for single *p-n* junction solar cells [20–24].

With such high economic potential from both the second- and third-generation solar cells, Si-based (first-generation) cells still produce the highest PV efficiency for terrestrial solar modules [25] and the most significant market share due to the well-established technology.

1.6.1 Operating Configuration of Photovoltaic Solar Cells

According to the literature [26–28], the two primary configurations of a thin-film solar cell are the substrate and superstrate device configurations. The classification depends on the sequence in which the layers are deposited (see Fig. 1.4).

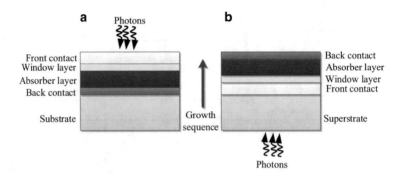

Fig. 1.4 Schematic cross section of substrate and superstrate configurations of thin-film solar cells

Both configurations are capable of generating high photon to electron conversion efficiency [24, 29–32]. A similar feature to both configurations is that photons enter the solar cell devices through the front contact and the window layer. But distinctively, in the superstrate configuration, photons pass through the glass before reaching the window layer/absorber layer junction. This is unlike the substrate configuration where photons are directly admitted to the window layer/absorber layer junction without any apparent or significant obstruction due to the shading of the cells by the front contacts. It is therefore crucial that the top transparent conducting oxide and glass utilised in the superstrate configuration must fulfil several stringent requirements, including low sheet resistance, temperature durability, excellent chemical stability, excellent adhesion and high optical transmission in the spectrum of interest.

The solar cell device fabrication work done during this programme as reported in this book uses the superstrate configuration. This is due to the following reasons:

1. The semiconductor deposition technique utilised: a conducting substrate such as fluorine-doped tin oxide (FTO) is required as the electrode on which the semiconductor is deposited using the electroplating technique (see Fig. 1.4b).
2. The metal back contact as required in substrate configuration has a high tendency of dissolving in the acidic aqueous electrolyte, thereby resulting in contamination/doping of the electrolyte and alteration of deposited material properties.
3. The cadmium telluride (CdTe) absorber layers utilised in this work have been known to have higher pinhole density when grown directly on transparent conducting oxide (TCO) such as FTO as compared to CdTe grown on cadmium sulphide (CdS) with minimum pinhole formation.

1.7 Photon Energy

Solar radiation comprises of elementary particles known as photons. A photon can be described as a discrete bundle (or quantum) of electromagnetic (or light) energy. A photon is characterised either by its wavelength (λ) or by its equivalent energy (E). Photon energy (E) is related to its frequency by the Equations 1.1 and 1.2.

$$E = hf \qquad \text{(Equation 1.1)}$$

$$E = \frac{hc}{\lambda} \qquad \text{(Equation 1.2)}$$

where E (J) is the photon energy, h is the Plank's constant given as 6.626×10^{-34} Js, f is the frequency measured in hertz (Hz), c is the speed of light given as 2.998×10^{8} ms^{-1} and λ is the wavelength (nm).

The relationship expressed by Equation 1.2 shows that light having low energy photons (such as "red" light) has long wavelengths while light having high energy photons (such as "blue" and "ultra violet" light) has short wavelengths.

Evaluation of the numerator expression in Equation 1.2 gives $hc = 1.99 \times 10^{-25}$ Jm. With the appropriate unit conversion, hc can also be written as:

$$hc = \left(1.99 \times 10^{-25}\,\text{Jm}\right) \times \left(\frac{1\,\text{eV}}{1.602 \times 10^{-19}\,\text{J}}\right) = 1.24 \times 10^{-6}\,\text{eVm}$$

Further, to convert the unit to nm (the units for λ),

$$\left(1.24 \times 10^{-6}\,\text{eVm}\right) \times \left(10^{9}\,\text{nm/m}\right) = 1240\ \text{eV nm}$$

Therefore, Equation 1.2 can be rewritten as:

$$E = \frac{1240}{\lambda}\ (\text{eV}) \qquad \text{(Equation 1.3)}$$

1.8   Photovoltaic Timeline and State of the Art

The photovoltaic effect was first observed in 1839 by Alexandre-Edmond Becquerel through experimentation with semiconductor materials. Other groups such as that of Daryl Chapin et al. from the Bell laboratories in 1954, Hoffman Electronics Corporation in 1960, etc. have all contributed to the development of PV solar technology. The increase in research and development in alternative energy generation technology was primarily due to the oil crisis in the 1970s. The importance of solar energy cannot be overemphasised, as its importance has been lauded in a scientific article as far back as 1911 with a catching caption, which reads "in the far distant future, natural fuels having been exhausted, 'solar power' will remain as the only means of existence of the human race" [33]. At present, the need for high-efficiency PV systems and reduction in the $W^{-1} cost is highly essential for the world's ever-growing population and demand for energy, to achieve sustainability. Towards achieving this task, a few of the landmarks by researchers and industries within the PV community are listed in Table 1.1.

Table 1.1 Timeline of photovoltaic solar energy technology [16]

Year	Events
1839	Discovery of PV by Edmund Becquerel when he was 19 years old
1883	Charles Fritts developed the first solar cell using elemental selenium as the light-absorbing material
1916	Robert Millikan experimentally proved photoelectric effect
1918	Jan Czochralski developed a method in which single crystal silicon can be grown
1923	Albert Einstein won the Nobel Prize for explaining the photoelectric effect
1954	4.5% efficient silicon solar cells were produced at Bell laboratory
1959	Hoffman Electronics produced 10% efficient silicon cells and was launched with PV array of 9600 cells
1960	Hoffman Electronics produced 14% efficient silicon solar cells
1970s	The first oil crisis gave a kick-start to search for low-cost alternative systems for terrestrial energy conversion, accelerating PV research activities
1980s	Thin-film CdTe and $CuInGaSe_2$ (CIGS) solar cells were introduced into the mainstream of PV research
1990s	Dye sensitised solar cell (DSSC) or "Grätzel solar cell" was introduced
2000s	Organic solar cells were introduced to the PV field
2001	CdTe-based solar panels of up to 0.94 m^2 with 10.4% efficiency
2013	First solar produced 16.1% efficiency for small area solar cell using thin-film CdTe
2014	First solar produced 20.4% efficiency for small area solar cell using thin-film CdTe
2015	First solar produced 21.5% efficiency for small area solar cell using thin-film CdTe
2016	First solar produced 22.1% efficiency for small area solar cell using thin-film CdTe

Other notable solar cell efficiencies documented in the literature include Si (crystalline) at 25.7%, gallium indium phosphide (GaInP) at 21.4%, copper indium gallium diselenide (CIGS) thin film at 22.6%, copper zinc tin sulphide selenium (CZTSS) thin film at 12.6%, copper zinc tin sulphide (CZTS) thin film at 11.0%, perovskite thin film at 22.1% and organic thin film at 12.1% [34] under one-sun illumination.

1.9   Research Aims and Objectives

The motivation for the work reported in this book is to advance the knowledge and technology of third-generation solar cells, through the development of research towards low-cost, high-efficiency electrodeposited devices. The work is based on a previously proposed and experimented model investigated by Dharmadasa in 2002 and 2005 [15, 35], which achieved a record conversion efficiency of 18% for a CdTe-based thin-film solar cell at the time. The main feature of the work reported by Dharmadasa's group was the n-n-heterojunction +large Schottky barrier configuration. The present work aims to incorporate a similar architecture and improve the conversion efficiency using graded bandgap device structures and low-cost electroplated (ED) semiconductor materials from aqueous solutions. The semiconductor materials explored in this book include ZnS (as the buffer layer), CdS (as the window layer) and CdTe (as the absorber layer), while the effect of in situ doping of CdTe with Cl, F, I and Ga was also investigated and reported [36–41]. Exploration of the semiconductor material involves the optimisation of the material layers through a study of their structural, compositional, morphological, optical and electrical properties. This was undertaken using facilities in the Material and Engineering Research Institute at Sheffield Hallam University (MERI-SHU). Other semiconductor materials utilised in this book were sourced within the research group and have been documented in the literature [35, 42–54] (see Table 1.2). The effect of post-growth treatment (PGT) using $CdCl_2$, $CdCl_2 + CdF_2$ and $CdCl_2 + Ga_2(SO_4)_3$ treatment on all the above properties of the electrodeposited layers and device performances of the fabricated solar cells were also explored and reported [55–57]. In this research programme, both the investigated and the outsourced semiconductor layers were incorporated into different graded bandgap configurations and reported [22, 24].

The distinct features of this research work include:

1. The use of thiourea ($SC(NH_2)_2$) instead of $Na_2S_2O_3$ as the sulphur (S) precursor for electrodeposited CdS to prevent sulphur precipitation and the accumulation of Na in the electrolytic bath [36]
2. The use of cadmium nitrate ($Cd(NO_3)_2$) instead of $CdSO_4$ as a precursor for CdTe due to the improved material and electronic quality of CdTe for the fabrication of CdS/CdTe solar cells
3. Incorporating $GaCl_3$ into the well-established $CdCl_2$ post-growth treatment [55–57], in situ doping of CdTe in an aqueous electrolytic bath [36–41]
4. The exploration of glass/FTO/n-CdS/n-CdTe/p-CdTe/Au configuration [24]

Figure 1.5 shows the outline of the work reported in this book, with the research objectives as follows:

1. Growth and optimisation of electrodeposited semiconductor materials (ZnS, CdS and CdTe) from aqueous electrolytic baths using two-electrode configuration.
2. Obtaining suitable deposition voltage range for the semiconductor layer deposition from cyclic voltammetric data.

    (a) Optimisation and study of layers performed through the study of the structural, compositional, morphological, optical and electrical properties of the ED-ZnS, ED-CdS and ED-CdTe layers using X-ray diffraction (XRD), Raman spectroscopy, scanning electron microscopy (SEM), UV-Vis spectroscopy and the photoelectrochemical (PEC) cell technique

3. Doping and study of the effect of in situ doping of CdTe with Cl, F, I and Ga.
4. Study of the effect of post-growth treatment (PGT) using $CdCl_2$, $CdCl_2 + CdF_2$ and $CdCl_2 + Ga_2(SO_4)_3$ treatment on the structural, optical, morphological properties and device performance of the fabricated solar cells.
5. Fabrication of solar cell devices incorporating the basic glass/FTO/n-CdS/n-CdTe heterojunction +large Schottky barrier (SB) at n-CdTe/metal interface

Table 1.2 Summary of explored electronic materials to date at author's research group using electroplating from aqueous solutions

Material electroplated	E_g (eV)	Precursors used for electroplating	Comments	References
CuInSe$_2$	~1.00	CuSO$_4$ for Cu ions, In$_2$(SO$_4$)$_3$ for In ions and H$_2$SeO$_3$ for Se ions	Ability to grow both p- and n-type material	[45]
CdTe	1.45	CdSO$_4$ or Cd(NO$_3$)$_2$ or CdCl$_2$ for Cd ions and TeO$_2$ for Te ions	Ability to grow both p- and n-type CdTe using Cd-sulphate, nitrate and chloride precursors	[42, 43]
CuInGaSe$_2$	1.00–1.70	CuSO$_4$ for Cu ions, In$_2$(SO$_4$)$_3$ for In ions, Ga$_2$(SO$_4$)$_3$ for Ga ions and H$_2$SeO$_3$ for Se ions	Ability to grow both p- and n-type material	[44]
CdSe	1.90	CdCl$_2$ for Cd ions and SeO$_2$ for Se ions	Work is in progress	[46]
InSe	1.90	InCl$_3$ for In ions and SeO$_2$ for Se ions	Work is in progress	[47]
GaSe	2.00	Ga$_2$(SO$_4$)$_2$ for Ga ions and SeO$_2$ for Se ions	Work is in progress	
ZnTe	1.90–2.60	ZnSO$_4$ for Zn ions and TeO$_2$ for Te ions	Ability to grow both p- and n-type material	[48]
CdS	2.42	CdCl$_2$ for Cd ions and Na$_2$S$_2$O$_3$, NH$_4$S$_2$O$_3$ or NH$_2$CSNH$_2$	Conductivity type is always n-type	[36, 49, 54]
CdMnTe	1.57–2.50	CdSO$_4$ for Cd ions, MnSO$_4$ for Mn ions and TeO$_2$ for Te ions	Work is in progress	
ZnSe	2.70	ZnSO$_4$ for Zn ions and SeO$_2$ for Se ions	Ability to grow both p- and n-type material	[50]
ZnO	3.30	Zn(NO$_3$)$_2$ for Zn ions		[51]
ZnS	3.75	ZnSO$_4$ for Zn and (NH$_4$)$_2$S$_2$O$_3$ for S ions	Ability to grow both p- and n-type material	[52]
Poly aniline (PAni)	–	C$_6$H$_5$NH$_2$ and H$_2$SO$_4$	To use as a pinhole plugging layer	[53]

and other configurations including glass/FTO/n-CdS/n-CdTe/p-CdTe/Au, glass/FTO/n-ZnS/n-CdS/n-CdTe/Au and glass/FTO/n-ZnS/n-CdS/n-CdTe/p-CdTe/Au were explored.

6. Assessment of the efficiency of the fabricated thin-film solar cells using current-voltage (I-V) measurement and developing these devices by optimisation of all processing steps to achieve highest possible efficiency. Device parameters were also assessed using capacitance-voltage (C-V) measurements.

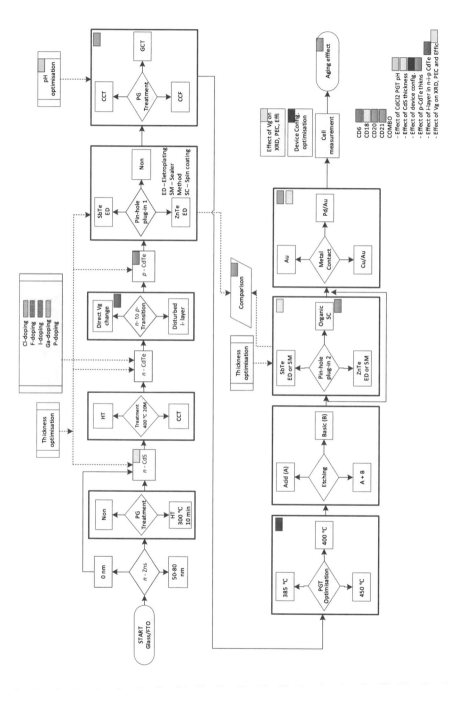

Fig. 1.5 Outline of the work reported in this book

1.10   Conclusions

This chapter presented in brief an overview of the need for solar energy research and an outline of the device characteristics of the range of solar cell devices being utilised and investigated to address the increasing demand for energy and the detrimental effect of conventional (non-renewable) energy sources. Amongst renewable energy sources, the enormity of solar energy, its origin and influence of air mass (AM) on the solar energy were discussed. A summary of the technology for harvesting solar energy with emphasis on photovoltaic solar cell and their timeline was also presented. The last section of this chapter presents the aims and objectives of this research programme focusing on next-generation solar cells.

References

1. J.P. Holdren, Population and the energy problem. Popul. Environ. **12**, 231–255 (1991). https://doi.org/10.1007/BF01357916
2. World population projected to reach 9.7 billion by 2050 | UN DESA | United Nations Department of Economic and Social Affairs. (n.d.). http://www.un.org/en/development/desa/news/population/2015-report.html. Accessed 9 Apr 2017
3. E.E. Michaelides, *Alternative Energy Sources* (Springer, Berlin, 2012). https://doi.org/10.1007/978-3-642-20951-2
4. A.M. Omer, Energy use and environmental impacts: a general review. J. Renew. Sustain. Energy. **1**, 53101 (2009). https://doi.org/10.1063/1.3220701
5. BP, BP Energy Outlook 2017. (2017), https://www.bp.com/content/dam/bp/pdf/energy-economics/energy-outlook-2017/bp-energy-outlook-2017.pdf. Accessed 9 Apr 2017
6. M. Dale, Meta-analysis of non-renewable energy resource estimates. Energy Policy **43**, 102–122 (2012). https://doi.org/10.1016/j.enpol.2011.12.039
7. N.S. Lewis, D.G. Nocera, Powering the planet: chemical challenges in solar energy utilization. Proc. Natl. Acad. Sci. U. S. A. **103**, 15729–15735 (2006). https://doi.org/10.1073/pnas.0603395103
8. NASA/Marshall Solar Physics. (n.d.), https://solarscience.msfc.nasa.gov/interior.shtml. Accessed 11 Apr 2017
9. D. Chiras, *Solar Electricity Basics: A Green Energy Guide.* (New Society Publishers, 2010), https://books.google.co.uk/books?id=_2brYQqb_RYC
10. O. Morton, Solar energy: Silicon Valley sunrise. Nature **443**, 19–22 (2006). https://doi.org/10.1038/443019a
11. Global Energy Budget | Precipitation Education. (n.d.), https://pmm.nasa.gov/education/lesson-plans/global-energy-budget. Accessed 25 Oct 2017
12. C.J. Riordan, Spectral solar irradiance models and data sets. Sol. Cells. **18**, 223–232 (1986). https://doi.org/10.1016/0379-6787(86)90121-3
13. The Greenhouse Effect and the Global Energy Budget | EARTH 103: Earth in the Future. (n.d.), https://www.e-education.psu.edu/earth103/node/1006. Accessed 25 Oct 2017
14. I.M. Dharmadasa, Third generation multi-layer tandem solar cells for achieving high conversion efficiencies. Sol. Energy Mater. Sol. Cells **85**, 293–300 (2005). https://doi.org/10.1016/j.solmat.2004.08.008
15. I.M. Dharmadasa, A.P. Samantilleke, N.B. Chaure, J. Young, New ways of developing glass/conducting glass/CdS/CdTe/metal thin-film solar cells based on a new model. Semicond. Sci. Technol. **17**, 1238–1248 (2002). https://doi.org/10.1088/0268-1242/17/12/306

16. I.M. Dharmadasa, *Advances in Thin-Film Solar Cells* (Pan Stanford, Singapore, 2013)
17. K. Masuko, M. Shigematsu, T. Hashiguchi, D. Fujishima, M. Kai, N. Yoshimura, T. Yamaguchi, Y. Ichihashi, T. Mishima, N. Matsubara, T. Yamanishi, T. Takahama, M. Taguchi, E. Maruyama, S. Okamoto, Achievement of more than 25% conversion efficiency with crystalline silicon heterojunction solar cell. IEEE J. Photovoltaics. **4**, 1433–1435 (2014). https://doi.org/10.1109/JPHOTOV.2014.2352151
18. First Solar raises bar for CdTe with 21.5% efficiency record: pv-magazine. (n.d.), http://www.pv-magazine.com/news/details/beitrag/first-solar-raises-bar-for-cdte-with-215-efficiency-record_100018069/#axzz3rzMESjUl. Accessed 20 Nov 2015
19. K. Ramanathan, M.A. Contreras, C.L. Perkins, S. Asher, F.S. Hasoon, J. Keane, D. Young, M. Romero, W. Metzger, R. Noufi, J. Ward, A. Duda, Properties of 19.2% efficiency ZnO/CdS/CuInGaSe2 thin-film solar cells. Prog. Photovolt. Res. Appl. **11**, 225–230 (2003). https://doi.org/10.1002/pip.494
20. W. Shockley, H.J. Queisser, Detailed balance limit of efficiency of p-n junction solar cells. J. Appl. Phys. **32**, 510 (1961). https://doi.org/10.1063/1.1736034
21. G. Conibeer, Third-generation photovoltaics. Mater. Today **10**, 42–50 (2007). https://doi.org/10.1016/S1369-7021(07)70278-X
22. I.M. Dharmadasa, A.A. Ojo, H.I. Salim, R. Dharmadasa, Next generation solar cells based on graded bandgap device structures utilising rod-type nano-materials. Energies **8**, 5440–5458 (2015). https://doi.org/10.3390/en8065440
23. M.A. Green, Third generation photovoltaics: ultra-high conversion efficiency at low cost. Prog. Photovolt. Res. Appl. **9**, 123–135 (2001). https://doi.org/10.1002/pip.360
24. A.A. Ojo, I.M. Dharmadasa, 15.3% efficient graded bandgap solar cells fabricated using electroplated CdS and CdTe thin films. Sol. Energy **136**, 10–14 (2016). https://doi.org/10.1016/j.solener.2016.06.067
25. M.A. Green, Solar cell efficiency tables (version 49). Prog. Photovolt. Res. Appl. **25**, 3–13 (2017). https://doi.org/10.1002/pip.2876
26. A. Romeo, M. Terheggen, D. Abou-Ras, D.L. Bätzner, F.-J. Haug, M. Kälin, D. Rudmann, A.N. Tiwari, Development of thin-film Cu(In,Ga)Se2 and CdTe solar cells. Prog. Photovolt. Res. Appl. **12**, 93–111 (2004). https://doi.org/10.1002/pip.527
27. B.E. McCandless, J.R. Sites, in *Handb. Photovolt. Sci. Eng.* Cadmium telluride solar cells (Wiley, Chichester, 2011), pp. 600–641. https://doi.org/10.1002/9780470974704.ch14.
28. T.L. Chu, S.S. Chu, Thin film II–VI photovoltaics. Solid State Electron. **38**, 533–549 (1995). https://doi.org/10.1016/0038-1101(94)00203-R
29. B.L. Williams, J.D. Major, L. Bowen, L. Phillips, G. Zoppi, I. Forbes, K. Durose, Challenges and prospects for developing CdS/CdTe substrate solar cells on Mo foils. Sol. Energy Mater. Sol. Cells **124**, 31–38 (2014). https://doi.org/10.1016/j.solmat.2014.01.017
30. A. Bosio, N. Romeo, S. Mazzamuto, V. Canevari, Polycrystalline CdTe thin films for photovoltaic applications. Prog. Cryst. Growth Charact. Mater. **52**, 247–279 (2006). https://doi.org/10.1016/j.pcrysgrow.2006.09.001
31. X. Wu, High-efficiency polycrystalline CdTe thin-film solar cells. Sol. Energy **77**, 803–814 (2004). https://doi.org/10.1016/j.solener.2004.06.006
32. L. Kranz, C. Gretener, J. Perrenoud, R. Schmitt, F. Pianezzi, F. La Mattina, P. Blösch, E. Cheah, A. Chirilă, C.M. Fella, H. Hagendorfer, T. Jäger, S. Nishiwaki, A.R. Uhl, S. Buecheler, A.N. Tiwari, Doping of polycrystalline CdTe for high-efficiency solar cells on flexible metal foil. Nat. Commun. **4**, 2306 (2013). https://doi.org/10.1038/ncomms3306
33. F. Shuman, Power from sunshine. Sci. Am. **105**, 291–292 (1911). https://doi.org/10.1038/scientificamerican09301911-291
34. M.A. Green, Y. Hishikawa, W. Warta, E.D. Dunlop, D.H. Levi, J. Hohl-Ebinger, A.W.H. Ho-Baillie, Solar cell efficiency tables (version 50). Prog. Photovolt. Res. Appl. **25**, 668–676 (2017). https://doi.org/10.1002/pip.2909
35. I. Dharmadasa, J. Roberts, G. Hill, Third generation multi-layer graded band gap solar cells for achieving high conversion efficiencies—II: experimental results. Sol. Energy Mater. Sol. Cells **88**, 413–422 (2005). https://doi.org/10.1016/j.solmat.2005.05.008

36. A.A. Ojo, I.M. Dharmadasa, Investigation of electronic quality of electrodeposited cadmium sulphide layers from thiourea precursor for use in large area electronics. Mater. Chem. Phys. **180**, 14–28 (2016). https://doi.org/10.1016/j.matchemphys.2016.05.006

37. H.I. Salim, O.I. Olusola, A.A. Ojo, K.A. Urasov, M.B. Dergacheva, I.M. Dharmadasa, Electrodeposition and characterisation of CdS thin films using thiourea precursor for application in solar cells. J. Mater. Sci. Mater. Electron. **27**, 6786–6799 (2016). https://doi.org/10.1007/s10854-016-4629-8

38. A.A. Ojo, H.I. Salim, O.I. Olusola, M.L. Madugu, I.M. Dharmadasa, Effect of thickness: a case study of electrodeposited CdS in CdS/CdTe based photovoltaic devices. J. Mater. Sci. Mater. Electron. **28**, 3254–3263 (2017). https://doi.org/10.1007/s10854-016-5916-0

39. A.A. Ojo, I.M. Dharmadasa, The effect of fluorine doping on the characteristic behaviour of CdTe. J. Electron. Mater. **45**, 5728–5738 (2016). https://doi.org/10.1007/s11664-016-4786-9

40. A.A. Ojo, I.M. Dharmadasa, Electrodeposition of fluorine-doped cadmium telluride for application in photovoltaic device fabrication. Mater. Res. Innov. **19**, 470–476 (2015). https://doi.org/10.1080/14328917.2015.1109215

41. A.A. Ojo, I.M. Dharmadasa, in *31st Eur. Photovolt. Sol. Energy Conf.* Effect of in-situ fluorine doping on electroplated cadmium telluride thin films for photovoltaic device application (2015), pp. 1249–1255. https://doi.org/10.4229/EUPVSEC20152015-3DV.1.40.

42. H.I. Salim, V. Patel, A. Abbas, J.M. Walls, I.M. Dharmadasa, Electrodeposition of CdTe thin films using nitrate precursor for applications in solar cells. J. Mater. Sci. Mater. Electron. **26**, 3119–3128 (2015). https://doi.org/10.1007/s10854-015-2805-x.

43. N.A. Abdul-Manaf, H.I. Salim, M.L. Madugu, O.I. Olusola, I.M. Dharmadasa, Electro-plating and characterisation of CdTe thin films using CdCl2 as the cadmium source. Energies **8**, 10883–10903 (2015). https://doi.org/10.3390/en81010883

44. I.M. Dharmadasa, N.B. Chaure, G.J. Tolan, A.P. Samantilleke, Development of p(+), p, i, n, and n(+)-type CuInGaSe2 layers for applications in graded bandgap multilayer thin-film solar. Cell **154**, 466–471 (2007). https://doi.org/10.1149/1.2718401.

45. I.M. Dharmadasa, R.P. Burton, M. Simmonds, Electrodeposition of CuInSe2 layers using a two-electrode system for applications in multi-layer graded bandgap solar cells. Sol. Energy Mater. Sol. Cells **90**, 2191–2200 (2006). https://doi.org/10.1016/j.solmat.2006.02.028

46. O.I. Olusola, O.K. Echendu, I.M. Dharmadasa, Development of CdSe thin films for application in electronic devices. J. Mater. Sci. Mater. Electron. **26**, 1066–1076 (2015). https://doi.org/10.1007/s10854-014-2506-x

47. M.L. Madugu, L. Bowen, O.K. Echendu, I.M. Dharmadasa, Preparation of indium selenide thin film by electrochemical technique. J. Mater. Sci. Mater. Electron. **25**, 3977–3983 (2014). https://doi.org/10.1007/s10854-014-2116-7

48. O.I. Olusola, M.L. Madugu, N.A. Abdul-Manaf, I.M. Dharmadasa, Growth and characterisation of n- and p-type ZnTe thin films for applications in electronic devices. Curr. Appl. Phys. **16**, 120–130 (2016). https://doi.org/10.1016/j.cap.2015.11.008

49. D.G. Diso, G.E.A. Muftah, V. Patel, I.M. Dharmadasa, Growth of CdS layers to develop all-electrodeposited CdS/CdTe thin-film solar cells. J. Electrochem. Soc. **157**, H647 (2010). https://doi.org/10.1149/1.3364800

50. A.P. Samantilleke, M.H. Boyle, J. Young, I.M. Dharmadasa, Electrodeposition of n-type and p-type ZnSe thin films for applications in large area optoelectronic devices. J. Mater. Sci. Mater. Electron. **9**, 231–235 (1998). https://doi.org/10.1023/A:1008886410204

51. J.S. Wellings, N.B. Chaure, S.N. Heavens, I.M. Dharmadasa, Growth and characterisation of electrodeposited ZnO thin films. Thin Solid Films **516**, 3893–3898 (2008). https://doi.org/10.1016/j.tsf.2007.07.156

52. M.L. Madugu, O.I.-O. Olusola, O.K. Echendu, B. Kadem, I.M. Dharmadasa, Intrinsic doping in electrodeposited ZnS thin films for application in large-area optoelectronic devices. J. Electron. Mater. **45**, 2710–2717 (2016). https://doi.org/10.1007/s11664-015-4310-7

53. N.A. Abdul-Manaf, O.K. Echendu, F. Fauzi, L. Bowen, I.M. Dharmadasa, Development of polyaniline using electrochemical technique for plugging pinholes in cadmium sulfide/cadmium

telluride solar cells. J. Electron. Mater. **43**, 4003–4010 (2014). https://doi.org/10.1007/s11664-014-3361-5

54. N.A. Abdul-Manaf, A.R. Weerasinghe, O.K. Echendu, I.M. Dharmadasa, Electro-plating and characterisation of cadmium sulphide thin films using ammonium thiosulphate as the sulphur source. J. Mater. Sci. Mater. Electron. **26**, 2418–2429 (2015). https://doi.org/10.1007/s10854-015-2700-5

55. A.A. Ojo, I.M. Dharmadasa, Optimisation of pH of cadmium chloride post-growth-treatment in processing CdS/CdTe based thin film solar cells. J. Mater. Sci. Mater. Electron. **28**, 7231–7242 (2017). https://doi.org/10.1007/s10854-017-6404-x

56. O.I. Olusola, M.L. Madugu, A.A. Ojo, I.M. Dharmadasa, Investigating the effect of GaCl3 incorporation into the usual CdCl2 treatment on CdTe-based solar cell device structures. Curr. Appl. Phys. **17**, 279–289 (2017). https://doi.org/10.1016/j.cap.2016.11.027

57. I.M. Dharmadasa, O.K. Echendu, F. Fauzi, N.A. Abdul-Manaf, O.I. Olusola, H.I. Salim, M.L. Madugu, A.A. Ojo, Improvement of composition of CdTe thin films during heat treatment in the presence of CdCl2. J. Mater. Sci. Mater. Electron. **28**, 2343–2352 (2017). https://doi.org/10.1007/s10854-016-5802-9

Chapter 2
Photovoltaic Solar Cells: Materials, Concepts and Devices

2.1 Introduction

This chapter focuses on a review of the literature and the science background of solar energy materials and solar cells. The various classifications of solid-state materials and the physics of junctions and interfaces in solar devices will be discussed. The main categories of solar cells will be presented in brief coupled with a general overview of next-generation solar cells.

2.2 Solid-State Materials

The prominent property peculiar to the classification of solid-state materials is their bandgap (E_g) as determined by the interatomic interaction resulting in valence band (E_v) and conduction band (E_c) energy states as defined by band theory [1]. Unlike the Bohr model of isolated atoms which exhibit discrete energy levels (or shells) and an electron configuration determined by atomic number [2], band theory defines the interaction between multiple atoms in which the discrete energy shells broaden into energy bands. The outermost shells of the atoms (with their various subshells) which are more loosely bound to respective nuclei merge to form more available energy levels within which the electrons can move freely. An increase in the number of atoms leads to the formation of discrete energy bands with energy levels that can be occupied by electrons. These allowed energy bands are separated by gaps in which there can be no electrons—known as band gaps, or forbidden energy gaps. The ease at which electrons can move between bands under the influence of excitation energy is determined by the value of the energy gap between the bands.

Figure 2.1a–c shows the schematic band diagrams of an electrical conductor, a semiconductor and an insulator, respectively. The conduction band (CB) is the electron-empty energy band and E_c is the lowest level of the CB, while the valence

© Springer International Publishing AG, part of Springer Nature 2019
A. A. Ojo et al., *Next Generation Multilayer Graded Bandgap Solar Cells*,
https://doi.org/10.1007/978-3-319-96667-0_2

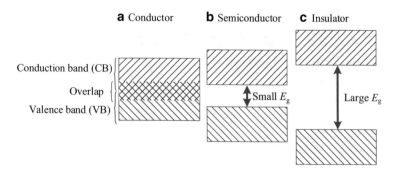

Fig. 2.1 Energy band diagrams of (**a**) a conductor, (**b**) a semiconductor and (**c**) an insulator

Table 2.1 Summary of main properties of different classes of solid-state materials

Parameter	Electrical conductors	Semiconductors	Electrical insulators
Electrical conductivity σ $(\Omega \, cm)^{-1} \equiv S$	$\sim 10^6 - 10^0$	$\sim 10^0 - 10^{-8}$	$\sim 10^{-8} - 10^{-20}$
Bandgap E_g (eV)	≤ 0.3	$\sim 0.3 - 4.0$	> 4.0

band (VB) is the allowed energy band that is filled with electrons at 0 K [1, 3] and the top of the VB is labelled E_v. For conductors (Fig. 2.1a) such as metals, the CB overlaps with the VB which is partially filled with electrons. Due to the overlap and the partially filled band, electrons move freely and require no external excitation to be promoted to the E_c [1]. Therefore the material possesses high conductivity attributable to the presence of conduction electrons contributing to current flow.

For both semiconductors and insulators, as respectively shown in Fig. 2.1b, c, their conduction bands are empty of electrons, valence bands are completely filled with electrons and there exists an energy bandgap of E_g between their E_v and E_c at 0 K [1, 3]. Due to the small energy gap between the E_c and E_v for semiconductors, an introduction of external excitation energy, such as via photons or thermal agitation at room temperature, can promote electrons from E_v to E_c leaving behind some unoccupied states in the valence band, known as holes. But for insulators, the bandgaps are large, making it difficult for electrons to be promoted from E_v to E_c. Therefore, the VB of an insulator is full of electrons and the CB is empty, limiting the number of charge carriers that are free to move and hence resulting in low electrical conductivity.

Further to the classification of solid-state material according to the energy bandgap (E_g), the electrical conductivity (σ) property can also be utilised [4, 5] (see Table 2.1).

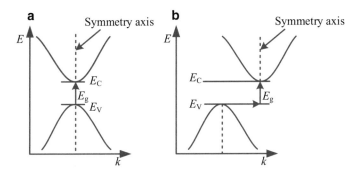

Fig. 2.2 Schematic plots of E-k diagrams for (**a**) direct bandgap semiconductor and (**b**) indirect bandgap semiconductor

2.2.1   Semiconductor Materials and Their Classification

Semiconductor materials are usually solid-state chemical elements or compounds with properties lying between that of a conductor and an insulator [3]. As shown in Table 2.1, they are often identified based on their electrical conductivity (σ) and bandgap (E_g) within the range of $\sim(10^0\text{--}10^{-8})$ $(\Omega\ \mathrm{cm})^{-1}$ and $\sim(0.3\text{--}4.0)$ eV, respectively [4]. Furthermore, semiconductor materials can also be classified based on their band alignment, elemental composition and dopant incorporation as respectively discussed in Sects. 2.2.1.1–2.2.1.3.

2.2.1.1   Classification Based on Band Symmetry

Classification of semiconductors can be based on the alignment of electron momentum (p) of the minimum energy difference between the bottom of the conduction band E_c and the top of the valence band E_v. Figure 2.2a, b shows the schematic diagrams of energy-momentum (E-k) plots for direct and indirect bandgap semiconductors, respectively.

Classically, the force on each charge carrier $F = m^*a$ (where m^* is the effective mass of electron or hole involved in the transition, a is the acceleration and v is the velocity). The momentum vector k of a charge carrier can be approximated from the kinetic energy E of the charge carrier as defined in Equation 2.1.

$$E = \tfrac{1}{2}m^*v^2 \qquad \text{(Equation 2.1)}$$
$$\text{where}\quad p = m^*v \qquad \text{(Equation 2.2)}$$

Therefore, Equation 2.1 can be rewritten as Equation 2.3

$$E(k) = \frac{p^2}{2m^*} \qquad \text{(Equation 2.3)}$$

or Equation 2.4 [3, 6] as redefined by de Broglie, where p equals $\hbar k$, $\hbar$ is the reduced Plank's constant defined as $\left(\frac{h}{2\pi}\right)$ and k is the wave vector which equals $\left(\frac{2\pi}{\lambda}\right)$.

$$E(k) = \frac{\hbar^2 k^2}{2m^*} \qquad \text{(Equation 2.4)}$$

For direct bandgap semiconductors, both the conduction band minima and the valence band maxima occur at the same crystal momentum. This implies that an electron at the top of the valence band can move to the bottom of the conduction band if it possesses sufficient energy, without any change in its momentum vector [3, 6]. Thus, an energised electron moves with a single effective mass (m^*) along the symmetry axis, and thereby momentum is conserved. Semiconductors in this category include ZnS, CdS, CdTe, etc. Contrarily, the conduction band minima and the valence band maxima occur at different crystal momenta for indirect bandgap semiconductor materials. This is consequential to a change in the momentum of the energised electron moving from the top of the valence band to the bottom of the conduction band. Thus, the involved energised electron will have two effective masses m_l^* and m_t^* which will respectively be longitudinal and transverse with respect to the symmetry axis, as shown in Fig. 2.2b. Phonons (a quantum of lattice vibration) which fundamentally possess a significant amount of momentum and relatively low energy make up for the difference in momentum in an electron energy transition in an indirect bandgap semiconductor [3, 6]. This participation of phonons is necessitated for the conservation of both energy and momentum for a fundamental transition of electron to be effected. Semiconductors in this category include Ge, Si, GaP, etc. Hence, for photonic processes, such as the photovoltaic effect, or light emission, a direct bandgap material is preferred due to the increased probability of an electron transition from the valence to conduction band or vice versa.

2.2.1.2 Classification Based on Elemental Composition

As documented in the literature, semiconductor materials utilised in photovoltaic applications are mostly crystalline or polycrystalline inorganic solids which lie between groups I and VI within the periodic table [4]. Based on elemental composition, semiconductor materials can be classified as elemental, binary, ternary or quaternary semiconductors (see Table 2.2).

Elemental semiconductors consist of a single element in group IV with typical examples including C, Si and Ge. Other compound semiconductor materials such as the binary (III–V and II–VI), ternary and quaternary semiconductors are produced when two, three or four elements chemically react with one another, respectively. In this research work, all the semiconductor materials grown and explored belong to the binary (II–VI) semiconductor group.

Table 2.2 Summary of semiconductor elements and compounds available for use in photovoltaic applications

Semiconductor family	Examples of semiconductors
Elemental semiconductors	C, Si, Ge
III–V semiconductors	AlN, AlP, AlAs, AlSb, GaN, GaP, GaAs, GaSb, InN, InP, InAs, InSb
II–VI semiconductors	ZnS, ZnSe, ZnTe, ZnO, CdS, CdSe, CdTe, CdO
Ternary compound semiconductors	$CuInSe_2$ (CIS), $Cd_xMn_{(1-x)}Te$ (CMT), $Cd_xHg_{(1-x)}Te$, $Al_xGa_{(1-x)}As$
Quaternary compound semiconductors	$CuInGaSe_2$ (CIGS), $AgInGaSSe_2$, $Cu_2ZnSnSSe_4$ (CZTS)

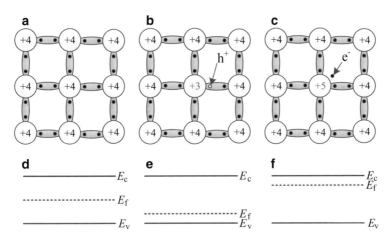

Fig. 2.3 Schematic diagram of (**a**) intrinsic, (**b**) p-doped and (**c**) n-doped semiconductor materials bonds (in Si). The band diagram of (**d**) intrinsic, (**e**) p-doped and (**f**) n-doped semiconductor materials (Si)

2.2.1.3 Classification Based on Dopants

Further to the classification of semiconductors based on band symmetry and elemental composition as discussed in Sects. 2.2.1.1 and 2.2.1.2, semiconductors can also be classified based on incorporated impurities: intrinsic and extrinsic semiconductors. Pure or undoped semiconductor materials without any significant incorporation of external dopant species are referred to as intrinsic or i-type semiconductors [6]. For example, an elemental semiconductor material such as silicon (Si) has four valence electrons in its outermost shell which are utilised in the formation of covalent bonds with other Si atoms as shown in Fig. 2.3a. Therefore, there are no free electrons in pure Si to partake in the flow of electric current. This results in the reduction of the electrical conductivity at absolute zero temperature. But with excitation energy equal or higher than the bandgap of the semiconductor, the only charge carriers are the electrons promoted to the E_c and the holes in the E_v that arise due to excitation of electrons to the E_c. Even in this state, the number of electrons in the E_c and the holes in the E_v are equal. For i-type semiconductor materials, the Fermi level is located in the middle of the bandgap as shown in Fig. 2.3d. The Fermi level defines the highest energy state within the bandgap that has a 50% probability of being occupied by electrons in a semiconductor material at any given time at absolute zero temperature. It should be noted that the electrical conduction type of compound semiconductor materials as discussed in Sect. 2.2.1.2 can either be dominated by intrinsic doping (based on the percentage composition of the elemental constituents [7]) or by intrinsic defect (resulting from Fermi level pinning [8]).

Extrinsic semiconductors are referred to as impure semiconductors due to the incorporation of external dopant element(s). The process or system of incorporating a suitable impurity into an intrinsic semiconductor in parts per million (ppm) level is referred to as doping. Depending on the included impurity, an extrinsic semiconductor can either be a p-type or an n-type semiconductor. It should be noted that the conductivity type of a semiconductor material is p-type provided holes are the majority carriers due to the inclusion of dopants from a group with lower valence electrons (acceptor impurity). And the conductivity type is n-type provided electrons are the majority carriers due to the inclusion of dopants from a group with higher valence electrons (donor impurity). As shown in Fig. 2.3b, c, doping Si which is a group IV element with a group III or group V element will result either in a p or n conductivity type due to the incorporation of excess holes or excess electrons, respectively [6]. Therefore, the Fermi level for the p-type material is positioned close to E_v (see Fig. 2.3e) and that of the n-type materials is positioned towards E_c (see Fig. 2.3f). In addition to the effect of dopants on the conductivity type of a semiconductor, native defects are also one of the principal factors which determine the Fermi level position in the semiconductor material [9, 10].

2.3 Junctions and Interfaces in Solar Cell Devices

Solar cell fabrication involves the formation of junctions between two or more semiconductors or between semiconductors and insulators or metals when brought in close contact with one another. As documented in the literature, the nature of the contact or junction formed is significant to the strength of the internal electric field, charge carrier creation and separation, pinning of the Fermi level and the formation of either an Ohmic or a rectifying (Schottky) contact. This section focuses on the types of the junctions formed, their properties and their applicability in solar cells.

2.3.1 Homojunction and Heterojunction

Junction formation in semiconductors can either be between layers of the same semiconductor material known as homojunction or between dissimilar semiconductor materials known as a heterojunction. Simple device configurations of both homojunction and heterojunction may be in the form of p^+-p, n-p or n-n^+ as demonstrated in the literature for solar cell device applications [6, 11–14]. Furthermore, depending on the doping concentrations of the semiconductor layers in contact, a junction can be considered as one sided or two sided.

2.3.2   p-n *and* p-i-n *Junction*

The *p-n* junction is regarded as the primary building block of most semiconductor application devices [3, 6]. Figure 2.4 shows the schematic illustration and energy band diagram of *p*- and *n*-type semiconductor materials prior to and after *p-n* junction formation. As shown in Fig. 2.4a, c, a *p-n* junction is formed between suitable *p*-type and *n*-type semiconductor materials.

Due to the excess of holes and electrons present in *p*-type and *n*-type semiconductor materials, respectively, when the semiconductors are in intimate contact,

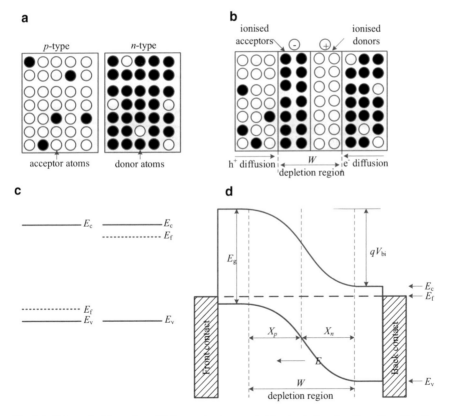

Fig. 2.4 Schematic illustration of (**a**) *p*- and *n*-type material prior to junction formation, (**b**) after *p-n* junction formation and energy band diagram of (**c**) *p*- and *n*-type semiconductor materials prior to the formation of *p-n* junction and (**d**) after close intimate contact formation

Fig. 2.5 Energy band
diagram of a *p-i-n* junction
device

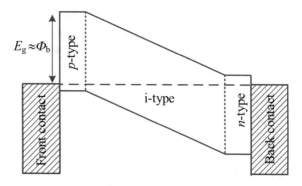

holes from the *p*-type material diffuse into the *n*-type material leaving behind
negatively charged acceptor atoms, while electrons from the *n*-type material diffuse
into the *p*-type material leaving behind positively charged donor atoms [3] (see
Fig. 2.4b). The diffusion leads to Fermi level equalisation and band bending as
shown in Fig. 2.4d. Owing to the presence of accumulated positive ion cores in the
n-type material and negative ion cores in the *p*-type material, an electric field *E* is
induced at the junction ($E = -dV/dx$ where *V* is the voltage and *x* is the distance
between plates), halting further diffusion of charge carriers. This region close to the
junction is referred to as the depletion region (*W*) or space-charge region. *W* is the
summation of the distances by which the depletion region extends into *p*-type (X_p)
and *n*-type (X_n) semiconductors, respectively. The values of X_p and X_n depend on the
doping concentration of the material. The supporting equations for these parameters
are later discussed in Sect. 3.5.2.

The *p-i-n* junction configuration as shown in Fig. 2.5 is a proceed of the *p-n*
junction with a sandwiched intrinsic (*i*-type) semiconductor layer in between *p*-
and the *n*-type layers. The functionality of a *p-i-n* diode is similar to that of a *p-n*
junction in which the Fermi levels of both the *p*- and the *n*-type semiconductors are
aligned through the *i*-type material. Due to the complexity of fabricating intrinsic
semiconductor materials [15], high resistive *p*- and *n*-type semiconductor materials
with low doping concentration may be utilised, since the depletion width is
dependent on the doping concentration of the semiconductor materials in contact.
The incorporated *i*-type semiconductor controls the depletion width depending on
its application [4, 16]. For applications such as photovoltaic devices, the incorporation of a wide *W* is essential for effective creation and separation of charge
carriers, but an optimisation of *W* is vital due to a reduction in the electric field;
$E = -dV/dx$.

Furthermore, the merit of this configuration includes the ability to achieve a
high potential barrier ϕ_b close to the bandgap of the semiconductor material [4]
which is synonymous to achieving high open-circuit voltage (V_{oc}). The advantages
of the *p-i-n* structure have been reported in the literature [16, 17].

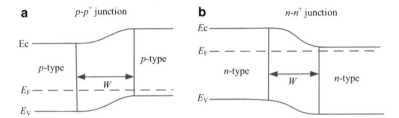

Fig. 2.6 Band diagrams of p-p^+ and n-n^+ junctions with smaller potential steps

2.3.3   p-p⁺ *and* n-n⁺ *Junction*

As mentioned in Sect. 2.3.1, the formation of p-p^+ or n-n^+ junctions may either be homojunction or heterojunction configuration depending on the semiconductor materials in contact. Figure 2.6 shows the band diagram of p-p^+ and n-n^+ junctions.

The observable characteristic of such junctions is a small potential step (low barrier height). The incorporation of such a configuration has been reported in the literature [13, 18] to give comparatively high photon to electron conversion.

2.3.4   *Metal-Semiconductor (M/S) Interfaces*

As previously discussed in Sect. 1.6, one of the prerequisites of effective photovoltaic energy conversion includes efficient transportation of the charge carriers through the external circuit [4]. For this to be achieved, at least two metal/semiconductor (M/S) contacts are required. Therefore selecting the appropriate metal contact with required M/S junction property is essential. Metal/semiconductor junctions can either be a non-rectifying (Ohmic) or a rectifying (Schottky) contact. Ohmic contact allows the flow of electric current in both directions across the M/S junction with the lowest resistance. The relationship between the current and voltage across the junction is linear (obeys Ohm's law). Conversely, a Schottky contact does allow current flow only in one direction across the junction [17, 19]. Notable parameters in the annotation of M/S contact includes the work function of the metal in contact (ϕ_m), the work function of the semiconductor (ϕ_s) and the electron affinity of the semiconductor (χ_s). The work function (ϕ) is defined as the minimum energy required to remove an electron to infinity from the surface of a given solid-state material. On the other hand, the electron affinity (χ) is defined as the amount of energy released or spent when an electron is added to a neutral atom [6].

2.3.4.1   Ohmic Contacts

Ohmic contacts are M/S contacts that possess negligible contact resistance relative to the bulk semiconductor resistance [6]. The formation of Ohmic contact is dependent on the relative energy difference (or barrier height ϕ_b) between the work function of

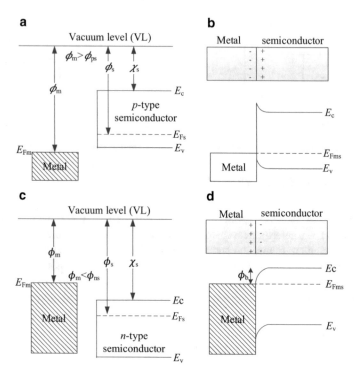

Fig. 2.7 Energy band diagrams for the formation of a Ohmic contact between a metal and a semiconductor. (**a**) Metal and *p*-type semiconductor before contact (**b**) after metal/semiconductor contact (**c**) metal and *n*-type semiconductor before contact and (**d**) after metal/semiconductor contact

the metal in contact (ϕ_m) and the electron affinity of the semiconductor (χ_s). With ϕ_b less than 0.4 eV [4], Ohmic contacts exhibit a narrow depletion region and negligible/non-rectifying capabilities. Figure 2.7a, b shows the energy band diagrams of a metal and *p*-type semiconductor before and after intimate contact where $\phi_m > \phi_s$, and Fig. 2.7c, d shows that of metal and *n*-type semiconductor where $\phi_m < \phi_s$.

For thermal equilibrium to be achieved between the *p*-type semiconductor and the metal as shown in Fig. 2.7a, b, electrons flow from the *p*-type semiconductor into the metal whereby increasing the hole concentration of the *p*-type semiconductor material [3]. Under forward-bias condition, holes produced in the *p*-type semiconductor can easily tunnel through into the metal from the semiconductor due to non-existence or minimal effect of the depletion region formed at the M/S junction.

On the contrary, for thermal equilibrium to be achieved between the *n*-type semiconductor and metal as shown in Fig. 2.7c, d, electrons flow from the metal to the semiconductor and increases the electron concentration of the *n*-type semiconductor [3]. Under forward-bias condition, electrons flow freely from the semiconductor to the metal due to the absence of any barrier. Under reverse bias condition, the electrons which flow from the metal to the semiconductor encounter a barrier $\phi_b = (E_c - E_{Fms})$ as shown in Fig. 2.7d. But due to the low barrier, electrons can still flow across. It should be noted that the analogy given above is under ideal conditions. Complexities due to surface states at the M/S junction or intrinsic defects [10, 20, 21] may cause the Fermi level to pin at an energy level whereby $V_{bi} > 0$ V, forming a rectifying contact instead of an Ohmic contact. This potential barrier height is then independent of the metal work function ϕ_m.

Alternatively, an Ohmic contact can be achieved by heavily doping the semiconductor material directly in contact with the metal [6]. The heavy doping of the semiconductor decreases the width of the depletion region, which increases the tunnelling probability of electrons through the barrier [6]. In other words, tunnelling becomes the dominant mechanism for current transport across the barrier, and this allows the flow of electrical current in both directions with linearity between current and bias voltage [1]. The current transport mechanism in M/S junctions will be further discussed in Sect. 2.3.4.3.

2.3.4.2 Rectifying (Schottky) Contacts

A rectifying contact is achieved between a semiconductor and a metal when the potential barrier height ($\phi_b = \phi_m - \chi_s$) is more than ~0.40 eV [4]. Consequently, a depletion region extending reasonably into the semiconductor is formed provided there exists a substantial difference between the work function of both the semiconductor (ϕ_s) and the metal (ϕ_m) in contact. The formation of a Schottky barrier between a p- and an n-type semiconductor with metal requires that $\phi_m < \phi_{sp}$ or $\phi_m > \phi_{sn}$ respectively which is unlike the Ohmic contact. Figure 2.8 shows the energy band diagrams for the formation of a Schottky barrier between a metal and either an n-type or a p-type semiconductor before and after metal/semiconductor contact. For this type of metal/n-type semiconductor contact, electrons flow from the n-type semiconductor to the metal due to the comparatively higher Fermi level of the semiconductor ($\phi_m > \phi_{sn}$). This lowers the n-type semiconductor Fermi level as a result of the reduction in electron concentration and bends the band until an alignment with the metal work function is reached and thermal equilibrium is established (Fig. 2.8b). Consequently, negative charges build up at the metal surface, and likewise positive charges build up in the semiconductor near the junction. This creates an internal electric field (E) and a depletion region of width (W) [3] around the M/S interface where band bending takes place. A similar phenomenon is experienced for Schottky M/S contact formed between a metal and a p-type semiconductor. In this case, electron flows from the metal to the semiconductor until thermal equilibrium is achieved as shown in Fig. 2.8c, d.

The built-in potential (V_{bi}) which prevents further diffusion of charge carriers across the depletion region and the barrier height (ϕ_b) of the metal/n-semiconductor Schottky contact can be mathematically defined as Equations 2.5 and 2.6, respectively.

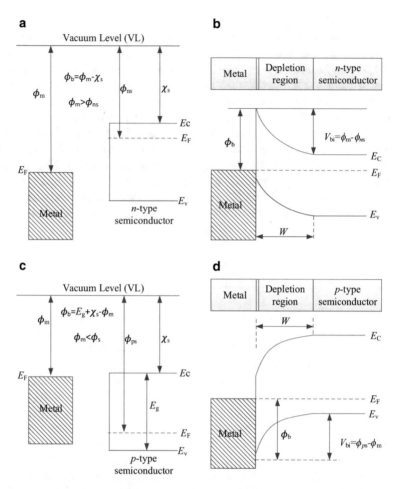

Fig. 2.8 Energy band diagrams for the formation of a Schottky barrier between a metal and a semiconductor. (**a**) Metal and *n*-type semiconductor before contact (**b**) after metal/semiconductor contact (**c**) metal and *p*-type semiconductor before contact and (**d**) after metal/semiconductor contact

$$V_{\mathrm{bi}} = (\phi_{\mathrm{m}} - \phi_{\mathrm{ns}}) \qquad\qquad \text{(Equation 2.5)}$$

$$\phi_{\mathrm{b}} = (\phi_{\mathrm{m}} - \chi_{\mathrm{s}}) \qquad\qquad \text{(Equation 2.6)}$$

The V_{bi} and the ϕ_{b} of the metal/*p*-semiconductor Schottky contact can be mathematically defined as Equations 2.7 and 2.8.

$$V_{\mathrm{bi}} = (\phi_{\mathrm{ps}} - \phi_{\mathrm{m}}) \qquad\qquad \text{(Equation 2.7)}$$

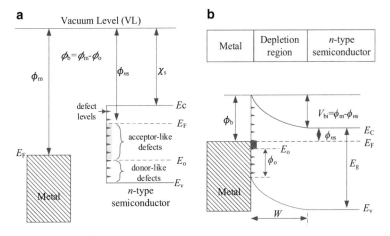

Fig. 2.9 Energy band diagrams for the formation of Schottky barrier between a metal and a semiconductor. (**a**) Metal and n-type semiconductor before contact (**b**) after metal/semiconductor contact incorporating surface states

$$\phi_b = \left(E_g + \chi_s - \phi_m\right) \qquad \text{(Equation 2.8)}$$

The ideal/theoretical narrative given above is quite different from practical applications due to complications associated with intrinsic defects, surface and interface states at the semiconductor/semiconductor (S/S) and M/S junctions [17, 20, 22]. Amongst such defects include interstitial and vacancy (dangling bond) [17] in the crystal lattice which are respectively due to transferred atoms from the surface into the interstitial site and incomplete bonding and lattice mismatch between consecutively grown semiconductor materials, impurities and oxide film formation at the interface [23] amongst others. Hence the dominant mechanism determines the pinning position of the Fermi level which may be independent of the metal work function ϕ_m [19, 22]. This observation has been documented in the literature for semiconductor materials such as CdTe, Cu(InGa)(SeS)$_2$ [10], GaAs, InP [24], Si [25], etc.

As documented in the literature, the defect region leads to the distribution of electronic levels within the forbidden bandgap at the interface as shown in Fig. 2.9a. The surface state is characterised by a neutral state E_o [17]. The states above the E_o contain acceptor-like defects which are neutral when empty but obtain electrons to become negative. The donor-like defect states below E_o are neutral when full, releasing an electron and becoming positive [6, 17, 23]. It should be noted that all energy states below E_F are occupied by electrons. Therefore, for a bare semiconductor, electrons accepted by the acceptor-like defects are taken from the semiconductor – just below the surface. This results in a band bending until equilibrium is reached between the interface state charge Q_{it} and the depletion region charge Q_D even prior to M/S contact [6]. Therefore Equation 2.6 can be rewritten as Equation 2.9 at high surface state density when E_F is pinned close to E_o.

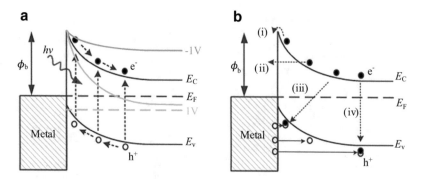

Fig. 2.10 Current transport mechanisms across a Schottky junction under (**a**) illuminated condition and biased by V_F and (**b**) dark condition and short circuited

$$\phi_b = E_g - \phi_o \qquad \text{(Equation 2.9)}$$

Therefore, it is essential that defects are adequately controlled due to their contribution in Schottky barrier formation which could be beneficial as in the case of impurity PV effect [26] or detrimental because they constitute of trap centres for charge carriers [27].

2.3.4.3   Current Transport Mechanisms Across Rectifying Contacts

Figure 2.10a, b shows the current transport mechanisms across a Schottky junction under illuminated and short-circuited condition and under forward bias (by V_F) in the dark condition, respectively.

The current transport mechanisms across a forward-biased Schottky barrier as shown in Fig. 2.10a includes:

1. Thermionic emission of electrons over the potential barrier of the semiconductor material into the metal. This is dominant for moderately doped semiconductor material with doping density $N \leq 10^{15}$ cm^{-3}.
2. Quantum mechanical tunnelling of electrons through the potential barrier. This is dominant for heavily doped semiconductor material which results in the thinning of the depletion region.
3. Recombination in the space-charge region.
4. Recombination in the neutral region due to hole injection from the metal into the semiconductor.

For an ideal Schottky barrier diode, thermionic emission (the thermally induced flow of charge carriers over the potential barrier) is the preferred current transportation mechanism, while mechanisms (2), (3) and (4) cause deviation from this ideal behaviour [6, 17, 19]. The effect of these current transport mechanisms is further elaborated in Sect. 3.5.1.1.

When the Schottky barrier solar cell is illuminated, the photo-generated electrons flow into the semiconductor. This direction is opposite to the electron flow under forward bias in dark condition. Therefore, when the I–V current is recorded under illuminated condition, the I–V curve shifts into the fourth quadrant since the photo-generated current is flowing in the opposite direction.

2.3.5 Metal-Insulator-Semiconductor (MIS) Interfaces

M/S Schottky devices are mainly characterised by comparatively lower barrier height (ϕ_b) and hence lower open-circuit voltage (V_{oc}) as compared to p-n junctions [6, 19]. The Schottky barrier represents approximately half of the p-n junction [4]. However, the barrier height of M/S junction can be increased close to the semiconductor bandgap by pinning the Fermi level close to the valence band and by the incorporation of an insulating layer in between the metal and the semiconductor interfaces as shown in Fig. 2.11. The insulating layer with an optimum thickness δ ranging between 1 and 3 nm [6] decouples the metal from the semiconductor as shown in Fig. 2.11. The incorporated i-layer eliminates the interface interaction between the metal and the semiconductor, thereby improving the lifetime of the electronic devices by reducing the degradation of the electrical contact and increasing both the band bending and the potential barrier height. On the other hand, the incorporation of the i-layer may result in the reduction of device efficiency due to a decrease in short-circuit current density (J_{sc}) and the FF as a consequence of the increase in the series resistance (R_s). The increase in the potential barrier height due to the insertion of the i-layer helps in increasing the V_{oc} of the device [6].

As reported in the literature, the i-layer may be an oxide layer [28, 29] and might not be grown intentionally [23]. The fabricated solar cell devices as reported in this book are mainly based on the Schottky barrier structure with no intentional oxide layer formation.

Fig. 2.11 Energy band diagram of an MIS interface showing enhancement in potential barrier height due to incorporation of thin insulating layer of thickness, δ, at the interface

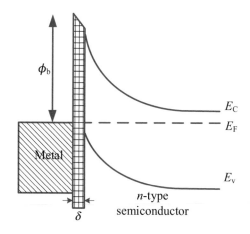

2.4 Types of Solar Cells

Although the basic functionality of a solar cell is photon to electron-hole (e-h) pair generation through the photovoltaic effect, which is a quantum mechanical phenomenon, the semiconductor material utilised (see Sect. 2.2.1) and the integrated technology are slightly different between devices. Hence, this section discusses in brief the underlining technology of different types of solar cells.

2.4.1 Inorganic Solar Cells

Inorganic materials such as silicon (Si) are by far the most researched, established and prevalent semiconductors in the photovoltaic market to date. So far, the highest efficiency solar cells have been fabricated using Si for terrestrial solar modules [30]. Inorganic semiconductor materials used in photovoltaic include mono- and polycrystalline silicon, amorphous silicon and other elemental, binary, ternary and quaternary compounds as discussed in Sect. 2.2.1.2 and shown in Table 2.2.

With crystalline silicon (c-Si) holding the dominant market share of photovoltaic cells, II–VI semiconductors such as cadmium telluride (CdTe) [31] are potentially one of the main rivals in terms of cost/watt. The main setback of CdTe is the toxicity of cadmium (Cd) and a limited supply of tellurium (Te). But as argued in the literature, Cd which is a by-product of zinc, lead and copper is more stable and less soluble in CdTe compound form, impossible to release Cd during normal CdTe cell operation and unlikely to dissociate into its constituents during fires in residential roofs [31]. Therefore utilisation of Cd in CdTe compound form reduces its toxicity while generating renewable energy. Aside from photovoltaic applications, CdTe has also been used as radiation detectors for X-rays, γ-rays, β-particles and α-particles [32]. Other applications also include detectors for nuclear medicine imaging [32].

The basic functionality of inorganic materials is the absorption of photons which breaks bonds between atoms and creates electron-hole pairs that are used for electricity generation.

2.4.2 Organic Solar Cells

Organic solar cells (OSC) are one of the developing technologies taking a niche at the dominance of Si-based solar cell due to properties such as reduction in the cost of production, the potential for mechanical flexibility of the solar module, reduced thickness of PV material and the fact that these devices can be fabricated using roll-to-roll production [33]. The main setback of OSCs is that they are still mainly under investigation and lab-based devices due to their long-term stability issues. The basic

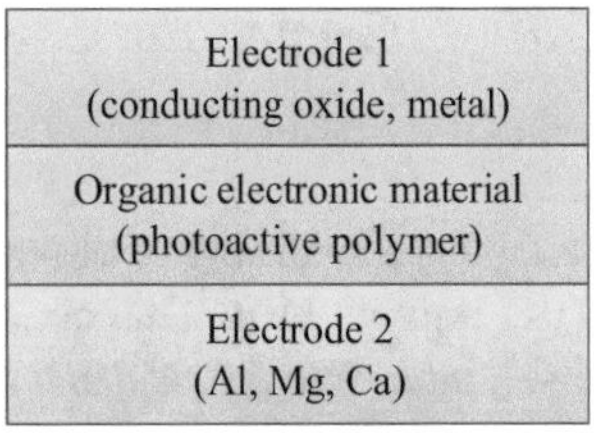

Fig. 2.12 Schematic
diagram of a single-layer
organic photovoltaic cell

| Electrode 1 (conducting oxide, metal) |
| Organic electronic material (photoactive polymer) |
| Electrode 2 (Al, Mg, Ca) |

technology requires the incorporation of photoactive polymer between metal contacts as shown in Fig. 2.12. Unlike the inorganic solar cells in which the electric fields are generated at the rectifying junctions which separate the e-h pairs created by the absorption of photons with energy higher than the semiconductor bandgap, the electric field generated in the organic solar cell is due to the differences in the work function ϕ between the two electrodes. When the organic polymer layer absorbs photons, electrons will be excited to the lowest unoccupied molecular orbital (LUMO) (equivalent to conduction band) and leave holes in the highest occupied molecular orbital (HOMO) (equivalent to valence band), thereby forming excitons [34]. The established potential difference due to the different work functions helps to split the excitons, pulling electrons to one electrical contact and holes to the other.

2.4.3   Hybrid Solar Cells

Hybrid solar cells also known as organic-inorganic solar cells are another potentially low-cost alternative to the well-established inorganic solar cells such as Si-based solar cells. The hybrid solar cell is a combination of both organic and inorganic materials, and therefore it combines the unique properties of highly efficient inorganic semiconductors with the low-cost film-forming properties of the conjugated polymers [35, 36]. The main disadvantages of the hybrid solar cell include numerous surface defects, improper organic-inorganic phase segregation, comparatively low charge mobility due to the disordered orientation of organic semiconductor molecules, instability issues at high relative humidity and lower efficiency as compared to established inorganic solar cells [37]. The basic materials utilised in the fabrication of hybrid solar cells include Si, cadmium compounds, metal oxide nanoparticles and low bandgap nanoparticles and carbon nanotubes (CNT) [38]. Examples of hybrid solar cells include the dye-sensitised solar cell (DSSC) and perovskite solar cell which differ from each other due to the light absorber layer incorporated. Based on the new light absorber material in the solid-state sensitised solar cell (perovskite), the power conversion efficiency (PCE) has increased to 22.1% from 11.9% in DSSC [39]. Once the main issue of instability and lifetime of hybrid solar cells are rectified, they are believed to have a high commercial potential to secure a substantial niche fraction of the photovoltaic market.

2.4.4 Graded Bandgap Solar Cells

The basic concept behind graded bandgap (GBG) solar cell is the possibility of effective harnessing of photons across the ultraviolet (UV), visible (Vis) and infrared (IR) regions [40]. This concept has been validated in the literature across organic [41], inorganic [26, 42] and hybrid [40, 41] solar cell technology. A graded bandgap can be achieved by grading the absorber layer in such a way that the bandgap varies throughout the entire thickness [26]. A GBG can also be produced by successively growing semiconductor materials on top of each other in which the layers are arranged such that the bandgap decreases gradually while the conductivity type gradually changes from one type to the other.

The earliest work that theoretically described the functionality of GBG configuration was reported by Tauc in 1957 [43]. His work elucidates the possibility of GBG solar cell configuration attaining higher conversion efficiency above the well-explored p-n junction cells and stimulated interest in the photovoltaic research community. Taking into consideration the photo-generated current, it was theoretically proved that GBG solar cell configurations are capable of attaining a conversion efficiency of ~38% under AM1.5 [44, 45] as compared to the 23% of single p-n junction solar cells. Based on this theoretical understanding, the first sets of graded bandgap cell were fabricated and reported in the 1970s [46, 47] by gradual doping of p-type gallium arsenide (GaAs) with aluminium. The published works demonstrated a single-sided p-type grading of $p\text{-}Ga_{1-x}Al_xAs/n\text{-}GaAs$ solar cell. In 2002, the first model of full solar cell device bandgap grading was proposed and published by Dharmadasa et al. [48]. The graded bandgap (GBG) architectures as proposed by Dharmadasa et al. [40, 49] can be facilitated by incorporating either an n-type or p-type wide bandgap front layer with a gradual reduction in bandgap towards p-type or n-type back layer respectively as shown in Fig. 2.13. The devices fabricated using the latter are more advantageous due to the higher potential barrier height achievable [42]. The advantages of graded bandgap solar cells include:

1. The possibility of harnessing photons across the solar spectrum (in the UV, Vis and IR region)

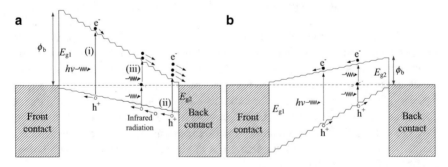

Fig. 2.13 Schematic representation of graded bandgap solar cells based on (**a**) p-type window layer and (**b**) n-type window layer [40, 41]

2. The reduction/elimination of thermalisation of "hot carriers" due to shared photon absorption at different regions of the solar cell
3. The improvement of e-h pair collection due to the presence of electric field (or depletion width) spanning approximately throughout the entire thickness of the solar cell to reduce recombination and generation (R&G)
4. Incorporation of impurity PV effect and impact ionisation in GBG solar cell configuration to reduce R&G and further increase photo-generated current density [26, 42]

Further to the underlying operating principles of photovoltaic solar cells as discussed in Sect. 1.6, the e-h pair generated due to the absorption of high energy photons as shown in Fig. 2.13a–i is effectively separated by the strong built-in electric field spanning approximately the width of the device. The strong built-in electric field owing to the steep gradient produced as a result of the device architecture forces the electron to accelerate towards the back contact. Such an electron moving at high kinetic energy (KE) may transfer its momentum to atoms located towards the rear end of the GBG device structure and break interatomic bonds, thereby creating additional e-h pairs as shown in Fig. 2.13a–i. This mechanism is referred to as band-to-band impact ionisation. On the other hand, the low energy photons near and within the IR region propagate towards the rear end of the GBG device structure. The IR can also be involved in the creation of e-h pairs either directly by exciting the electron from the valence band (E_v) to the conduction band (E_c) or predominantly by promoting the electrons to the defect levels within the forbidden gap. The presence of defects at the M/S junction as previously discussed in Sect. 2.3.4 can be positively utilised if adequately controlled [26]. The defects present may have been the naturally existing defects within the material. The recombination of the e-h pairs can be avoided due to the immediate transfer of the electron's paired hole as a result of the existing strong electric field in the GBG device. Therefore, the electron becomes trapped in the defect level until further absorption of an IR photon to promote the electron to the conduction band. The newly excited electron can push the initially excited electrons at one of the defect levels to a higher defect level or to the E_c. This mechanism is known as impurity PV effect [26, 50] (see Fig. 2.13a–iii). Further to this, there is a possibility of having both impact ionisation and impurity PV effect mechanism function in unison. This is achieved by the transfer of momentum from high KE electron accelerating to the GBG device rear end to promote the trapped electron in the defect level during impurity PV effect. The combination of both impact ionisation and impurity PV effect results in an avalanche effect during this process.

The downside to bandgap grading in GBG solar cells includes high technicality required in growth and the unpredictability of layer characteristic after post-growth treatment amongst others.

Table 2.3 Summary of the time-scale CdS/CdTe solar cell efficiency landmarks

Year	η (%)	V_{oc} (mV)	J_{sc} (mAcm^{-2})	FF	Team
1977	8.5	–	–	–	Matsushita
1981	8.9	710	16.5	0.58	Kodak
1983	10.3	725	21.2	0.67	Monosolar
1988	11.0	763	20.1	0.72	AMETEK
1990	12.3	783	25.0	0.63	Photon energy
1991	13.4	840	21.9	0.73	UoSF
1992	14.6	850	24.4	0.71	UoSF
1993	15.8	843	25.1	0.75	UoSF
1997	16.0	840	26.1	0.73	Matsushita
2001	16.4	848	25.9	0.75	NREL
2001	16.5	845	25.9	0.76	NREL
2002	10.6	823	20.4	0.63	BP Solar
2011	17.3	842	29.0	0.76	First Solar
2012	18.3	857	27.0	0.77	GE
2013	18.7	852	28.6	0.77	First Solar
2013	19.6	857	28.6	0.80	GE
2014	21.4	876	30.3	0.79	First Solar
2016	22.1	887	31.7	0.79	First Solar

2.5   Next-Generation Solar Cell Overview

The competitiveness of solar cell technology lies within its conversion efficiency, while its economic viability is determined by its production cost. Semiconductor deposition techniques such as electrodeposition matches the economic requirements provided efficient solar cells can be fabricated using the technique. Improvements of CdS/CdTe solar cells after about two decades of stagnation have been achieved through better understanding of materials and device issues. In view of increasing efficiency and/or economic viability, several concepts and innovative technologies proposed include intermediate band solar cell [44], quantum dots and quantum wells solar cell [51], down- and upconversion solar cell [52], plasmonic solar cell [53], hot-carrier solar cell [54], tandem solar cell with tunnel junctions [55], dye-sensitised and Perovskite solar cells [56] and graded bandgap solar cell [49, 57]. Amongst the proposed concepts, only tandem solar cells and graded bandgap solar cells have experimentally shown reasonable efficiency with required stability and the potential for increasing performance. The main difference between these two approaches is that in the tandem solar cells, efficiency increase is mostly due to an increase in open-circuit voltage, V_{oc}, while in the graded bandgap (GBG) solar cells, the efficiency is mainly increased due to increase in the short-circuit current density, J_{sc}. The high J_{sc} in GBG is a result of E_c to E_c and E_v to E_v connectivity (parallel connection) along the GBG interfaces [49]. The main drawback of tandem solar cells is the series connection of cells resulting into the reduction in charge carrier mobility and increase in the recombination of e-h pairs generated within the cell, while the graded bandgap solar cell's disadvantage is the technicality of growth required.

2.6 CdS/CdTe-Based Solar Cells

Based on electronic and optoelectronic properties such as a near-ideal direct bandgap of 1.45 eV, a high absorption coefficient $>10^4$ cm^{-1} at 300 K and the ability to absorb all usable energy from the solar spectrum within a thickness <2 μm [58], CdTe has proved to be an excellent II–VI semiconductor material for PV applications. CdTe-based solar cells have been well explored to develop low-cost and high-efficiency solar cells as an alternative to the present-day fossil fuel-dependent energy sources which are harmful to our ecosystem sustainability.

The exploration of CdTe-based solar cells is dated as far back as 1947 with the measurement of photoconductivity of incomplete phosphors of CdTe [59]. The photovoltaic potential of the CdTe-based solar cell was first theoretically demonstrated by Loferski in 1956 [60]. His work iterated that semiconductor materials with bandgaps close to 1.5 eV would possess the highest possible conversion efficiencies. Adirovich et al. published their first work on CdS/CdTe polycrystalline solar cell with superstrate configuration in 1970 [58, 61] with PV conversion efficiency $>2\%$. Based on the extensiveness of the research work undergone within the CdTe-based PV field afterward, the listing as shown in Table 2.3 can only be viewed as a selection of the landmarks in CdTe-based solar cells till date.

2.7 Conclusions

This chapter discussed in brief the properties of solid-state materials with main emphasis on semiconductors and their classifications. The properties of semiconductor which makes them suitable for photovoltaic applications and properties of the junctions formed between S/S and M/S were also discussed. Further to this, the basic operations, advantages and disadvantages of different solar cell types were also presented in addition to the brief overview of next-generation solar cells.

References

1. J. Singh, *Electronic and Optoelectronic Properties of Semiconductor Structures* (Cambridge University Press, Cambridge, 2003). https://doi.org/10.1017/CBO9780511805745
2. I.A. Sukhoivanov, I.V. Guryev, *Photonic Crystals* (Springer, Berlin, 2009). https://doi.org/10.1007/978-3-642-02646-1
3. D. Neamen, Semiconductor physics and devices. Mater. Today. **9**, 57 (2006). https://doi.org/10.1016/S1369-7021(06)71498-5
4. I.M. Dharmadasa, *Advances in Thin-Film Solar Cells* (Pan Stanford, Singapore, 2013)
5. W.H. Strehlow, E.L. Cook, Compilation of energy band gaps in elemental and binary compound semiconductors and insulators. J. Phys. Chem. Ref. Data. **2**, 163–200 (1973). https://doi.org/10.1063/1.3253115

6. S.M. Sze, K.K. Ng, *Physics of Semiconductor Devices* (Wiley, Hoboken, 2006). https://doi.org/10.1002/0470068329
7. A.A. Ojo, I.M. Dharmadasa, 15.3% efficient graded bandgap solar cells fabricated using electroplated CdS and CdTe thin films. Sol. Energy. **136**, 10–14 (2016). https://doi.org/10.1016/j.solener.2016.06.067
8. S.D. Sathaye, A.P.B. Sinha, Studies on thin films of cadmium sulphide prepared by a chemical deposition method. Thin Solid Films **37**, 15–23 (1976). https://doi.org/10.1016/0040-6090(76)90531-9
9. I.M. Dharmadasa, J.M. Thornton, R.H. Williams, Effects of surface treatments on Schottky barrier formation at metal/n-type CdTe contacts. Appl. Phys. Lett. **54**, 137 (1989). https://doi.org/10.1063/1.101208
10. I.M. Dharmadasa, J.D. Bunning, A.P. Samantilleke, T. Shen, Effects of multi-defects at metal/semiconductor interfaces on electrical properties and their influence on stability and lifetime of thin film solar cells. Sol. Energy Mater. Sol. Cells. **86**, 373–384 (2005). https://doi.org/10.1016/j.solmat.2004.08.009
11. T.L. Chu, S.S. Chu, C. Ferekides, J. Britt, C.Q. Wu, Thin-film junctions of cadmium telluride by metalorganic chemical vapor deposition. J. Appl. Phys. **71**, 3870–3876 (1992). https://doi.org/10.1063/1.350852
12. L. Huang, Y. Zhao, D. Cai, Homojunction and heterojunction based on CdTe polycrystalline thin films. Mater. Lett. **63**, 2082–2084 (2009). https://doi.org/10.1016/j.matlet.2009.06.028
13. B.E. McCandless, J.R. Sites, in *Handb. Photovolt. Sci. Eng.* Cadmium telluride solar cells (Wiley, Chichester, 2011), pp. 600–641. https://doi.org/10.1002/9780470974704.ch14.
14. M.P. Mikhailova, A.N. Titkov, Type II heterojunctions in the GaInAsSb/GaSb system. Semicond. Sci. Technol. **9**, 1279–1295 (1994). https://doi.org/10.1088/0268-1242/9/7/001.
15. P. Hofmann, *Solid State Physics: An Introduction*, 2nd edn. (Wiley-VCH, Berlin, 2015)
16. P.V. Meyers, Advances in CdTe n-i-p photovoltaics. Sol. Cells. **27**, 91–98 (1989). https://doi.org/10.1016/0379-6787(89)90019-7
17. E.H. Rhoderick, The physics of Schottky barriers? Rev. Phys. Technol. **1**, 81–95 (1970). https://doi.org/10.1088/0034-6683/1/2/302
18. W.G. Oldham, A.G. Milnes, n-n Semiconductor heterojunctions. Solid. State. Electron. **6**, 121–132 (1963). https://doi.org/10.1016/0038-1101(63)90005-4
19. E.H. Rhoderick, Metal-semiconductor contacts. IEE Proc. I Solid State Electron Devices. **129**, 1 (1982). https://doi.org/10.1049/ip-i-1.1982.0001
20. J.P. Ponpon, A review of ohmic and rectifying contacts on cadmium telluride. Solid. State. Electron. **28**, 689–706 (1985). https://doi.org/10.1016/0038-1101(85)90019-X
21. I.M. Dharmadasa, Recent developments and progress on electrical contacts to CdTe, CdS and ZnSe with special reference to barrier contacts to CdTe. Prog. Cryst. Growth Charact. Mater. **36**, 249–290 (1998). https://doi.org/10.1016/S0960-8974(98)00010-2
22. J. Bardeen, Surface states and rectification at a metal semi-conductor contact. Phys. Rev. **71**, 717–727 (1947). https://doi.org/10.1103/PhysRev.71.717
23. J. Singh, *Semiconductor Devices: Basic Principles* (Wiley, New York, 2001)
24. W.E. Spicer, I. Lindau, P. Skeath, C.Y. Su, P. Chye, Unified mechanism for Schottky-barrier formation and III-V oxide interface states. Phys. Rev. Lett. **44**, 420–423 (1980). https://doi.org/10.1103/PhysRevLett.44.420
25. R. Schlaf, R. Hinogami, M. Fujitani, S. Yae, Y. Nakato, Fermi level pinning on HF etched silicon surfaces investigated by photoelectron spectroscopy. J. Vac. Sci. Technol. A **17**, 164 (1999). https://doi.org/10.1116/1.581568
26. I.M. Dharmadasa, O. Elsherif, G.J. Tolan, Solar cells active in complete darkness. J. Phys. Conf. Ser. **286**, 12041 (2011). https://doi.org/10.1088/1742-6596/286/1/012041
27. H.J. Queisser, Defects in semiconductors: some fatal, some vital. Science **281**, 945–950 (1998). https://doi.org/10.1126/science.281.5379.945
28. R.B. Godfrey, M.A. Green, Enhancement of MIS solar-cell "efficiency" by peripheral collection. Appl. Phys. Lett. **31**, 705–707 (1977). https://doi.org/10.1063/1.89487

29. W.A. Nevin, G.A. Chamberlain, Effect of oxide thickness on the properties of metal-insulator-organic semiconductor photovoltaic cells. IEEE Trans. Electron Devices. **40**, 75–81 (1993). https://doi.org/10.1109/16.249427

30. M.A. Green, Solar cell efficiency tables (version 49). Prog. Photovolt. Res. Appl. **25**, 3–13 (2017). https://doi.org/10.1002/pip.2876

31. V.M. Fthenakis, Life cycle impact analysis of cadmium in CdTe PV production. Renew. Sustain. Energy Rev. **8**, 303–334 (2004). https://doi.org/10.1016/j.rser.2003.12.001

32. B.M. Basol, High-efficiency electroplated heterojunction solar cell. J. Appl. Phys. **55**, 601–603 (1984). https://doi.org/10.1063/1.333073

33. J. Nelson, Polymer:fullerene bulk heterojunction solar cells. Mater. Today. **14**, 462–470 (2011). https://doi.org/10.1016/S1369-7021(11)70210-3

34. J. Nelson, Organic photovoltaic films. Curr. Opin. Solid State Mater. Sci. **6**, 87–95 (2002). https://doi.org/10.1016/S1359-0286(02)00006-2

35. S. Gunes, N.S. Sariciftci, Hybrid solar cells. Inorganica Chim. Acta. **361**, 581–588 (2008). https://doi.org/10.1016/j.ica.2007.06.042

36. M. Wright, A. Uddin, Organic-inorganic hybrid solar cells: a comparative review. Sol. Energy Mater. Sol. Cells. **107**, 87–111 (2012). https://doi.org/10.1016/j.solmat.2012.07.006

37. W. Xu, F. Tan, X. Liu, W. Zhang, S. Qu, Z. Wang, Z. Wang, Efficient organic/inorganic hybrid solar cell integrating polymer nanowires and inorganic nanotetrapods. Nanoscale Res. Lett. **12**, 11 (2017). https://doi.org/10.1186/s11671-016-1795-9

38. P.-L. Ong, I.A. Levitsky, Organic/IV, III-V semiconductor hybrid solar cells. Energies. **3**, 313–334 (2010). https://doi.org/10.3390/en3030313

39. NREL efficiency chart. (n.d.), https://www.nrel.gov/pv/assets/images/efficiency-chart.png. Accessed 19 June 2017

40. I.M. Dharmadasa, Third generation multi-layer tandem solar cells for achieving high conversion efficiencies. Sol. Energy Mater. Sol. Cells. **85**, 293–300 (2005). https://doi.org/10.1016/j.solmat.2004.08.008

41. O. Ergen, S.M. Gilbert, T. Pham, S.J. Turner, M.T.Z. Tan, M.A. Worsley, A. Zettl, Graded bandgap perovskite solar cells. Nat. Mater. **16**, 522–525 (2016). https://doi.org/10.1038/nmat4795

42. I.M. Dharmadasa, A.A. Ojo, H.I. Salim, R. Dharmadasa, Next generation solar cells based on graded bandgap device structures utilising rod-type nano-materials. Energies. **8**, 5440–5458 (2015). https://doi.org/10.3390/en8065440

43. J. Tauc, Generation of an emf in semiconductors with nonequilibrium current carrier concentrations. Rev. Mod. Phys. **29**, 308–324 (1957). https://doi.org/10.1103/RevModPhys.29.308

44. M. Wolf, Limitations and possibilities for improvement of photovoltaic solar energy converters: part I: considerations for earth's surface operation. Proc. IRE. **48**, 1246–1263 (1960). https://doi.org/10.1109/JRPROC.1960.287647

45. P.R. Emtage, Electrical conduction and the photovoltaic effect in semiconductors with position-dependent band gaps. J. Appl. Phys. **33**, 1950–1960 (1962). https://doi.org/10.1063/1.1728874

46. M. Konagai, K. Takahashi, Graded-band-gap pGa1-xAlxAs-nGaAs heterojunction solar cells. J. Appl. Phys. **46**, 3542–3546 (1975). https://doi.org/10.1063/1.322083

47. H.J. Hovel, J.M. Woodall, Ga[sub 1−x]Al[sub x]As-GaAs P-P-N heterojunction solar cells. J. Electrochem. Soc. **120**, 1246 (1973). https://doi.org/10.1149/1.2403671

48. I.M. Dharmadasa, A.P. Samantilleke, N.B. Chaure, J. Young, New ways of developing glass/conducting glass/CdS/CdTe/metal thin-film solar cells based on a new model. Semicond. Sci. Technol. **17**, 1238–1248 (2002). https://doi.org/10.1088/0268-1242/17/12/306

49. I. Dharmadasa, J. Roberts, G. Hill, Third generation multi-layer graded band gap solar cells for achieving high conversion efficiencies—II: experimental results. Sol. Energy Mater. Sol. Cells. **88**, 413–422 (2005). https://doi.org/10.1016/j.solmat.2005.05.008

50. A.S. Brown, M.A. Green, Impurity photovoltaic effect: fundamental energy conversion efficiency limits. J. Appl. Phys. **92**, 1329–1336 (2002). https://doi.org/10.1063/1.1492016

51. K.W.J. Barnham, G. Duggan, A new approach to high-efficiency multi-band-gap solar cells. J. Appl. Phys. **67**, 3490–3493 (1990). https://doi.org/10.1063/1.345339
52. T. Trupke, M.A. Green, P. Würfel, Improving solar cell efficiencies by down-conversion of high-energy photons. J. Appl. Phys. **92**, 1668 (2002)
53. Y.Y. Lee, W.J. Ho, Y.T. Chen, Performance of plasmonic silicon solar cells using indium nanoparticles deposited on a patterned TiO2 matrix. Thin Solid Films. **570**, 194–199 (2014). https://doi.org/10.1016/j.tsf.2014.05.022
54. Y. Takeda, T. Motohiro, Highly efficient solar cells using hot carriers generated by two-step excitation. Sol. Energy Mater. Sol. Cells. **95**, 2638–2644 (2011). https://doi.org/10.1016/j.solmat.2011.05.023
55. J.F. Geisz, D.J. Friedman, J.S. Ward, A. Duda, W.J. Olavarria, T.E. Moriarty, J.T. Kiehl, M.J. Romero, A.G. Norman, K.M. Jones, 40.8% efficient inverted triple-junction solar cell with two independently metamorphic junctions. Appl. Phys. Lett. **93**, 123505 (2008). https://doi.org/10.1063/1.2988497
56. A.B.F. Martinson, M.S. Góes, F. Fabregat-Santiago, J. Bisquert, M.J. Pellin, J.T. Hupp, Electron transport in dye-sensitized solar cells based on ZnO nanotubes: evidence for highly efficient charge collection and exceptionally rapid dynamics. J. Phys. Chem. A. **113**, 4015–4021 (2009). https://doi.org/10.1021/jp810406q
57. T. Stelzner, M. Pietsch, G. Andrä, F. Falk, E. Ose, S. Christiansen, Silicon nanowire-based solar cells. Nanotechnology. **19**, 295203 (2008). http://stacks.iop.org/0957-4484/19/i=29/a=295203
58. F.V. Wald, Applications of CdTe. A review. Rev. Phys. Appliquée. **12**, 277–290 (1977). https://doi.org/10.1051/rphysap:01977001202027700
59. R. Frerichs, The photo-conductivity of "incomplete phosphors". Phys. Rev. **72**, 594–601 (1947). https://doi.org/10.1103/PhysRev.72.594
60. J.J. Loferski, Theoretical considerations governing the choice of the optimum semiconductor for photovoltaic solar energy conversion. J. Appl. Phys. **27**, 777–784 (1956). https://doi.org/10.1063/1.1722483
61. A. Luque, S. Hegedus, *Handbook of Photovoltaic Science and Engineering* (Wiley, Chichester, 2010). https://doi.org/10.1002/9780470974704

Chapter 3
Techniques Utilised in Materials Growth and Materials and Device Characterisation

3.1 Introduction

Most semiconductor deposition techniques are capable of producing high-quality thin films using complex systems at a very high manufacturing cost. However, new research is directed at reducing material usage and cost of production and improving the fabricated device performances. Consequently, the cost of many devices including solar cells has been reduced by using thin films [1, 2]. The deposition of thin-film semiconductor materials with thicknesses ranging from a few nanometres (nm) to micrometers (μm) is achievable using several techniques as documented in the literature. The main deposition technique being focused on in this program is electrodeposition due to its significance in the research work as reported in this book. Further to this deposition method, selected analytical methods utilised in characterising and optimising the deposited semiconductor materials and devices will also be discussed.

3.2 Overview of Thin-Film Semiconductor Deposition Techniques

Thin-film semiconductor deposition techniques can be broadly aligned under physical or chemical deposition categories. The physical deposition refers to the technologies in which material is released from a source and deposited on a substrate using thermodynamic, electromechanical or mechanical processes [3, 4]. This includes closed-space sublimation (CSS), molecular beam epitaxy (MBE), radio frequency (RF) sputtering and magnetron sputtering amongst others. On the other hand, the chemical deposition techniques are accomplished by the utilisation of precursors either in their liquid or gaseous state to produce a chemical reaction on the surface of a substrate, leaving behind chemically deposited thin-film coatings on the

© Springer International Publishing AG, part of Springer Nature 2019
A. A. Ojo et al., *Next Generation Multilayer Graded Bandgap Solar Cells*,
https://doi.org/10.1007/978-3-319-96667-0_3

substrate. Such technique includes electrodeposition (ED), chemical bath deposition (CBD), chemical reduction plating, spin coating and chemical vapour deposition (CVD) processes. Other subdivisions of chemical vapour deposition (CVD) include metalorganic CVD (MOCVD), atmospheric pressure CVD (APCVD), etc. [3].

However, based on the simplicity of technique, columnar growth, self-purification, ease of extrinsic doping [5] and ease of depositing n-type, i-type and p-type semiconductors (for semiconductor materials whose conductivity type depends on stoichiometry) by varying the deposition potential, scalability, manufacturability [6–8] and economic viability [9] amongst other advantages, electrode-position is considered as one of the leading thin-film deposition techniques [10]. Electrodeposition is not without its challenges; this includes requirements for a conducting substrate, pinhole generation (due to its nucleation mechanism [11]), low process temperature (as improved crystallinity is achievable at higher deposition temperature [12]), comparatively longer deposition time, etc.

3.3 Electrodeposition Growth Technique

Electrodeposition is the process of depositing metals or compound semiconductors on a conducting substrate by passing an electric current through an ionic electrolyte in which metal or semiconductor ions are inherent [13]. The passage of current is required due to the inability of the chemical reaction resulting in the deposition of the solid material on the conducting substrate to proceed on its own as a result of positive free energy change ΔG of the reaction.

Although specific electrodeposition methods can be categorised based on the power supply source used, the working electrode and electrode configuration (as shown in Fig. 3.1), the basic deposition mechanism and setup remains similar across these variations. The basic deposition mechanism entails the flow of electrons from the power supply to the cathode. Positively charged cations are attracted towards the cathode and negatively charged anions to the anode. The cations or anions are neutralised electrically by gaining electrons (through a reduction process) or losing electrons (through an oxidation process) and being deposited on the working electrode (WE), respectively [13]. The typical ED setup of two-electrode (2E) configuration as shown in Fig. 3.2a consists of deposition container (beakers), deposition electrolyte, magnetic stirrer, hotplate, power supply, a working electrode, a counter electrode and an optional reference electrode (RE) in the case of three-electrode (3E) configuration (see Fig. 3.2b).

In this research work, a potentiostatic power source and cathodic deposition in which semiconductors are deposited on the cathode in a two-electrode (2E) config-uration were utilised in the deposition of all the semiconductor layers. The use of potentiostatic power source was due to the effect of deposition voltage on the atomic percentage composition of elements in the electrodeposited layer, which is one of the factors determining the conductivity type [14, 15]. Cathodic deposition was utilised due to its ability to produce stoichiometric thin films with good adherence to the

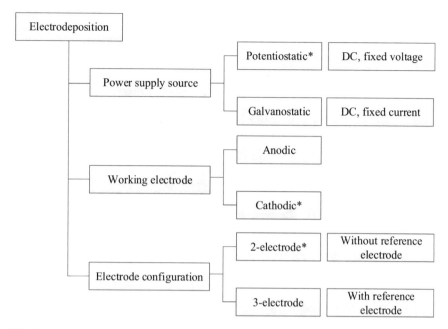

Fig. 3.1 The main categories of electrodeposition technique. * signifies the options utilised in this research work

substrate as compared to anodic deposition [16]. In conjunction, most of the metal ions deposited as reported in this book are cations.

The two-electrode configuration as shown in Fig. 3.2a was utilised due to its industrial applicability and process simplification and also to eliminate possible Ag^+ and K^+ ions doping [17, 18] which may emerge from the Ag/AgCl or saturated calomel electrode (SCE) reference electrodes (see Fig. 3.2b). Taking the electrode-position of n-CdS and n-CdTe layers which are respectively utilised as the main window and absorber layers in this work into perspective, both K^+ and Ag^+ from group I of the periodic table are considered as p-type dopants. Therefore, any leakage of K^+ and Ag^+ into the electrolytic bath may result in compensation leading to the growth of highly resistive material which has a detrimental effect on the efficiency of fabricated solar cells. This has been experimentally shown and reported in the literature [17].

The two-electrode electrodeposition configurations are not without challenges, with the main challenge being the fluctuation or drop in the potential measured across the cathode and the anode during deposition. This is due to the alteration in resistivity of the substrate with increasing semiconductor layer thickness and the change in the ionic concentration of the electrolyte. This differs from the three-electrode configuration, where the potential difference is measured across the working and the reference electrodes while the measured current is between the working and the counter electrodes. In general, other factors such as the pH of the electrolyte [19], applied deposition potential [14, 15], deposition temperature [20], stirring rate

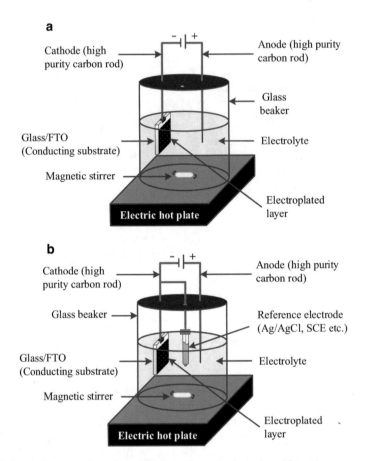

Fig. 3.2 Typical (**a**) two-electrode and (**b**) three-electrode electrodeposition set up

[21] and concentration of ions in the deposition electrolyte [19] affect the electro-deposition process and the properties of the deposited layers.

The electrodeposition of both metals and semiconductors is governed by Faraday's laws of electrodeposition. The first law states that "The amount of substance liberated or deposited at an electrode is directly proportional to the quantity of electricity passed through the electrolyte." This principle can be mathematically represented by Equation 3.1, where m is the mass of a substance liberated or deposited on an electrode, Q is the total electric charge and Z is a constant known as equivalent chemical weight:

$$m = ZQ \qquad \text{(Equation 3.1)}$$

Faraday's second law of electrolysis states that, when the same quantity of electricity is passed through several electrolytes, the mass of the substances deposited or liberated are proportional to their respective chemical equivalent or equivalent weight. The chemical equivalent weight can be defined as the ratio of molecular weight to the valence number of ions (electrons transferred per ion) as shown in Equation 3.2, where M is the molecular mass (gmol^{-1}), z is the valence number, n is the number of electrons transferred in the chemical reaction for the formation of 1 mole of substance in gcm^{-3} and F is the Faraday's constant (96,485 Cmol^{-1}):

$$Z = \left(\frac{M}{z}\right) = \left(\frac{M}{nF}\right)$$
(Equation 3.2)

Therefore, Equation 3.1 can be rewritten as Equation 3.3:

$$m = \left(\frac{Q}{F}\right)\left(\frac{M}{n}\right)$$
(Equation 3.3)

knowing that:

$$Q = it$$
(Equation 3.4)

$$\rho = \frac{m}{TA}$$
(Equation 3.5)

where i is the average deposition current (A), t is the deposition time (s), J is current density (Acm^{-2}), ρ is the density (gcm^{-3}), T is the thickness (cm) and A is the surface area of the substrate in contact with the electrolyte (cm^2). Therefore, Equation 3.3 can be further modified as Equation 3.6:

$$T = \left(\frac{1}{nF}\right)\left(\frac{itM}{\rho A}\right) = \left(\frac{JtM}{nF\rho}\right)$$
(Equation 3.6)

The relationship can be used to estimate the thickness of electrodeposited layers.

3.4 Material Characterisation Techniques

3.4.1 Cyclic Voltammetry

Cyclic voltammetry is a technique utilised in the study of the electrochemical reaction of the electrodeposition process by measuring the electric current through the electrolyte as a function of potential sweep across the electrodes [22]. In other words, it can be defined as the current-voltage characteristics of the electrolyte that describes the conduction of electrical current through the electrolyte (resulting in the oxidation or reduction of ions) as a result of the potential applied. Therefore it is necessary to perform a voltammetric study on electrolytic baths from which semi-conductor layers are to be deposited to identify the cathodic voltage range suitable for their deposition.

Cyclic voltammetry is performed by scanning the potential across both the working and counter electrodes in the forward and reverse direction to qualitatively identify both the deposition and dissolution sequence of elements of the semiconductor. With this qualitative information, the suitable region from which the deposition of near-stoichiometric semiconductor layers can be determined for further characterisation using techniques will be discussed in the following sections.

For the cyclic voltammetric study in this work, the power source utilised is Gill AC computerised potentiostat (ACM instrument), while the voltage range scanned and scan rate were 0–2000 mV and 3 mVs^{-1}, respectively.

To further pinpoint the deposition potential (V_i) in which stoichiometric semiconductor material can be deposited based on material properties, characterisation of the deposited layers within the observed deposition range from the voltammetric study needs to be explored. The structural features of the deposited layers were obtained using the X-ray diffraction (XRD) and Raman spectroscopy. The morphological features were obtained using scanning electron microscopy (SEM). The compositional analyses were performed using energy-dispersive X-ray spectroscopy (EDX). The optical features were determined using ultraviolet-visible (UV-Vis) spectroscopy. The electronic property such as conductivity type was determined using photoelectrochemical (PEC) cell measurement, while the resistivity was determined using the direct current (DC) conductivity measurement technique.

3.4.2   X-Ray Diffraction (XRD) Technique

X-ray diffraction (XRD) is a non-contact and non-destructive technique utilised in the determination of the structural properties of materials. Information on lattice parameters and sample texture such as atomic planes, phase identification, lattice spacing, intensity of individual diffractions, preferred orientation and crystallite size can be obtained.

Figure 3.3 shows the main components of XRD equipment which comprises of the sample holder, X-ray tube and X-ray detector. The sample to be investigated is placed in the sample holder. X-rays generated in the X-ray tube are channelled through a monochromator (to select radiation of a single wavelength or energy) to be filtered and then assembled using a collimator (to produce a parallel beam of rays or radiation). The X-rays are further channelled through both the divergent and anti-scatter slits to limit the divergence of the incident X-ray beam and output noise (due to amorphous or air scattering), respectively. The interaction between the X-rays after being channelled through the monochromator and the slits onto the sample under investigation is referred to as interference. During XRD operation, the incident angle of the X-rays on the sample under investigation changes due to the movement of the X-ray tube along the diffractometer circle to focus the rays onto the sample under investigation.

The interaction between X-rays and atoms of a crystal results in the elastic scattering of the rays by the electrons of the atoms. The reflection of incoming

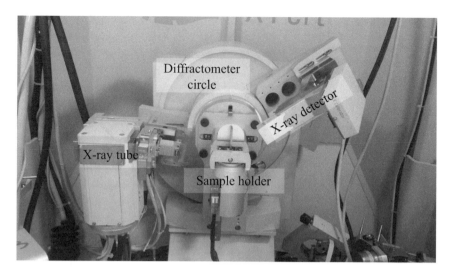

Fig. 3.3 XRD equipment setup showing its main components

rays by the electrons in the atoms gives rise to diffracted rays. It should be noted that X-rays are electromagnetic radiation within the wavelength range of (1–100) Å. The X-ray wavelength range falls within the same order of magnitude as the typical interatomic distance in a crystal. For this reason, it is easy for X-rays to be diffracted by the crystal structures [23]. Scattered rays may either support one another (constructive interference) or interfere and eliminate one another (destructive interference). It should be noted that the principle of XRD is based on constructive interference which is governed by Bragg's law.

Therefore, the conditions of interaction as stated by Bragg's law shown in Equation 3.7 is fulfilled when the interaction between the sample being investigated and the incident X-rays having wavelength λ (Å) produces constructive interference and diffracted rays:

$$n\lambda = 2d \sin \theta \qquad \text{(Equation 3.7)}$$

where n is an integer, λ is the wavelength, d is the lattice spacing (interatomic distance) and θ is the angle between the incident beam and atomic plane. The Bragg's law is derived from Fig. 3.4, where the path difference between the two adjacent X-rays ABC and EFG is PF + FQ which is equal to $2d \sin \theta$.

Therefore, if $2d \sin \theta$ is equal to the integral multiple of the X-ray wavelength ($n\lambda$), the result is constructive interference which produces a peak in intensity. Then the X-ray detector detects, processes and converts the diffracted X-rays signal to a count rate which is sent to a computerised device as a function of the diffraction angle 2θ. The data obtained using XRD includes d-spacing for diffraction which can be compared with the Joint Committee on Powder Diffraction Standards (JCPDS) for easy identification of crystalline elements and compounds.

Fig. 3.4 The basic principle of X-ray diffraction showing constructive interference when Bragg's law is satisfied

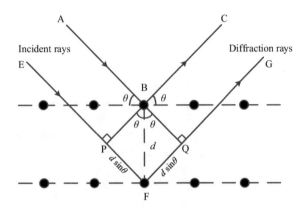

Based on the XRD peaks obtained, crystallite size can be evaluated using Debye-Scherrer's equation as shown in Equation 3.8. Where D is the crystallite size (nm), λ is the X-ray wavelength (Å), β is the full width at half maximum (rad) of the investigated peak, θ is the Bragg's angle (°) and the constant $K = 0.94$ is the geometry of spherical crystal. However, it should be noted that there is a limitation of the use of Scherrer's equation in determining the crystallite size. This equation is formulated to calculate grain sizes within few tens of nanometres [24]:

$$D = \frac{K\lambda}{\beta \cos \theta} \qquad \text{(Equation 3.8)}$$

In this work, the X-ray diffraction (XRD) scans were carried out within the range of $2\theta = (20\text{--}70)°$ using Philips PW X'Pert diffractometer with Cu-K$_\alpha$ monochromator having a wavelength of 1.54 Å. The source tension and current were set to 40 kV and 40 mA, respectively. Between $2\theta = (20\text{--}70)°$, a step count of ~2000 was utilised amounting to a possible uncertainty range of ±0.025°. The obtained diffraction peaks were fitted and analysed using PANalytical X'Pert HighScore Plus software package.

3.4.3 Raman Spectroscopy Technique

Raman spectroscopy is a very efficient and non-destructive characterisation technique which can be utilised for the identification and investigation of molecular fingerprint, crystallinity, strain and stress of the solid-state materials [25]. The technique relies on the inelastic scattering of monochromatic light, which is mainly laser source ranging from ultraviolet (244–380) nm through visible (380–750) nm to near infrared (750–900) nm.

As shown in Fig. 3.5, the interaction between the laser light and the molecules of the material under investigation results in the vibration of the molecules which eventually leads to the emission of scattered light and when returned to the

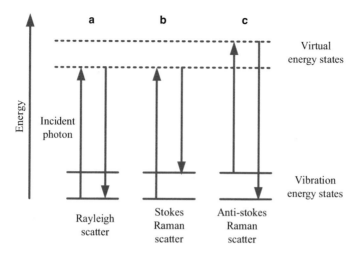

Fig. 3.5 Schematic representation of the Raman transitional schemes: (**a**) Rayleigh scattering, (**b**) Stokes-Raman scattering and (**c**) anti-Stokes-Raman scattering

molecules returns to the vibration energy states after excitation. The change in vibration energy state between the original state and the new state gives the shift in the energy of the emitted photon [26]. The excited molecule may return to its original vibration energy state and emit scattered radiation with frequency (v_f) equal to the frequency of the incident laser light in the case of Rayleigh scattering. Molecules may return to a higher or lower vibration energy state than the original state, and this phenomenon is known as Stokes-Raman scattering or anti-Stokes-Raman scattering, respectively. The emitted scattered light is detected by a photon detector, analysed in computerised Raman equipment and displayed as Raman spectrum on the computer visual display unit. The Raman spectra reported in this book were obtained using Renishaw's Raman spectrometer with an argon laser and an excitation wavelength of 514 nm.

3.4.4 Scanning Electron Microscopy (SEM) Technique

Scanning electron microscopy (SEM) is a technique utilised for morphological analysis of material surfaces. Figure 3.6 shows a schematic diagram of SEM equipment with the inclusion of its major components. Scanning electron microscopy is performed in a high vacuum chamber (of about 10^{-4} Nm^{-2}) in which the surface of the material under investigation is scanned with a focused beam of high-energy electrons generated at high voltage in the electron gun. The electron gun situated at the upper part of the microscope column may contain either a tungsten filament, lanthanum hexaboride (LaB_6) or cerium hexaboride (CeB_6) solid-state

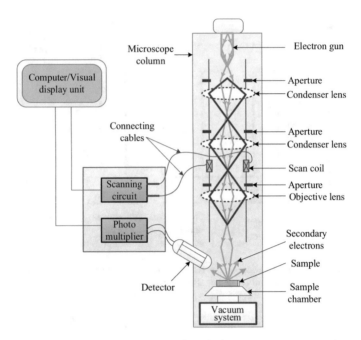

Fig. 3.6 Schematic diagram of a scanning electron microscope showing its main components

crystal sources and a field emission gun (FEG) which are the main sources of electrons in an SEM equipment [27].

Electrons are produced at the electron source at a high voltage ranging between 5 and 30 kV depending on the equipment specifications. The generated beam of high-energy electrons is channelled down the microscope column towards the sample being investigated through a series of condenser lenses and apertures. The condenser lenses focus the incident electron beam on to the sample under investigation, while the aperture is the opening through which the electron beam is channelled. The positioning of the electron beam on the sample under investigation is controlled by scan coils placed above the objective, lens as shown in Fig. 3.6. The scan coil also allows the electron beam to scan the sample surface area in a raster scan pattern [28]. The interaction between the primary electron beam and the atoms in the sample under investigation produces several signals based on both inelastic and elastic interactions [28]. Resulting signals include secondary electrons, backscattered electrons, Auger electrons and characteristic X-rays [28] which contain information about the surface topography and composition sample under investigation. The resulting signals are sensed and collected by the detectors and processed to produce images visible on the visual display unit. As shown in Fig. 3.7, the probing depth of the signals resulting from these interactions is different within the sample from which they can escape due to their unique physical properties and energies [28].

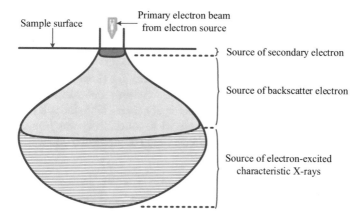

Sample surface

Primary electron beam
from electron source

} Source of secondary electron

Source of backscatter electron

Source of electron-excited
characteristic X-rays

Fig. 3.7 Typical diagram illustrating the production of three different signals within the specimen volume as a result of sample-electron interaction

Image formation in SEM is accomplished by mapping either the intensity of secondary electron (SE) or the backscattered electron (BSE) signals from the specimen [28]. The secondary electrons (SE) are low-energy electron ejected from the outer shells of atoms in the sample under investigation after the inelastic interaction between the primary electron beam and the sample. Secondary electrons are created close to the surface of the sample within a depth of (5–50) nm [28] and therefore produce information on the morphology of the sample under investigation. SE are more influenced by surface properties (such as sample work function and local surface curvature) rather than by the atomic number. It should be noted that an increase in atomic number increases the SE signal produced [29]. The SE signals are detected by a secondary electron detector (SED).

Backscattered electrons (BSE) are higher-energy electrons which originate from the elastic interaction between the primary electron source and the nuclei of atoms within the sample. The interaction results in minimal energy loss and scattering of the electrons in a different direction. The electrons scattered towards the surface of the sample are detected by the backscattered electron detector (BSED). The BSE gives compositional information on the samples. The intensity of BSE signal is affected by the atomic number of the specimen, with higher-atomic-numbered element showing higher intensities; this is because atoms having higher atomic numbers easily backscatter [28, 29]. BSE occur at a notable depth within the sample depending on the beam energy and the sample composition [28].

The SEM images reported in this book were obtained using FEI Nova 200 NanoSEM at a magnification of ×60,000 and an accelerating voltage of 10 kV. For comparability, the accelerating voltage for all the layers explored was kept constant due to the relationship as shown in Equation 3.9 [30], where A is the atomic weight, H is the penetration depth, V is the acceleration voltage, z is the atomic number and ρ is the density. Based on Equation 3.9, the probing depth for CdS and CdTe are ~955 nm and ~879 nm, respectively:

$$H = \frac{0.0276AV^{1.67}}{z^{0.89}\rho}$$ (Equation 3.9)

The investigated samples (glass/FTO/deposited layer) were thoroughly cleaned using methanol, rinsed in deionised water and dried in a stream of nitrogen gas. The sample was attached to the conductive SEM stub using conductive carbon adhesive discs. Silver paint was also utilised to electrically connect the stub and glass/FTO on which the thin-film was deposited. The electrical connectivity was required to avoid noise in SEM image due to charging effects which occur on non-conductive material as in this case, the glass/FTO substrate.

3.4.5   Energy-Dispersive X-Ray (EDX) Technique

Energy-dispersive X-ray spectroscopy (EDX) is an analytical technique utilised for the compositional analysis of elements contained in a material. It is a technique used in conjunction with SEM as discussed in Sect. 3.4.4. The resulting signals due to the interaction between the primary electron beam and sample under investigation include secondary electrons, backscattered electrons, Auger electrons and characteristic X-rays [28]. The SEM image utilises secondary electrons and backscattered electrons for image formation, while the EDX uses the characteristic X-rays for elemental composition analysis.

Characteristic X-rays are generated by inelastic interactions between the primary electron beam and the inner shell electrons of the atom of the sample under investigation. The inner shell electron is emitted from the sample's atom (ionisation) [28] as shown in Fig. 3.8. The hole created in the inner shell is filled by the electrons from the outer shell in a process known as relaxation of an atom into its neutral state.

The transition of an electron from the outer shell to the inner shell during relaxation releases a characteristic amount of energy from the atom in the form of X-ray photon which is equivalent to the energy difference between the two shells.

Fig. 3.8 Schematic diagram of energy-dispersive X-ray spectroscopy (EDX)

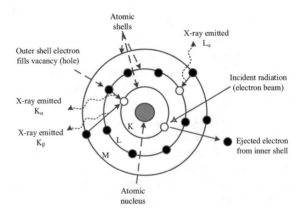

The characteristic X-ray is a fingerprint of the element from which it is emitted. The emitted X-ray is sensed and collected by the EDX detector. The EDX detector also separates the characteristic X-rays of the different elements into different energy spectra. As shown in Fig. 3.8, the nomenclature of the emitted X-rays is dependent on the inner shell on which the hole is situated and the outer shell from which the electron transit to fill the hole. The EDX measurements reported in this book were carried out using EDX detector attached to FEI Nova 200 NanoSEM.

It should be noted that the downside to the use of EDX includes:

1. Limitation of EDX sensitivity of element(s) to ~1000 ppm by weight unless counting time is increased
2. Analytical accuracy of ~2% due to uncertainties due to the composition of the standard utilised, overlapping elemental peaks and background corrections [31]

Due to this possible error range of ±2%, the elemental concentration data obtained were processed as a ratio one element to the other to nullify the error.

3.4.6 Ultraviolet-Visible (UV-Vis) Spectrophotometry Technique

The UV-Vis spectroscopy is a technique used for the analysis of the optical properties of an optical material. Such optical properties include the absorbance, transmittance and reflectance of light incident on an optical material. The technique is performed using a UV-Vis spectrophotometer which is capable of scanning wavelength range between the near ultraviolet (UV) ~(200–380) nm and visible (Vis) regions ~(380–750) nm.

As shown in Fig. 3.9, the basic components of a UV-Vis spectrophotometer comprise of a light source, monochromator, photodetector, signal processor and

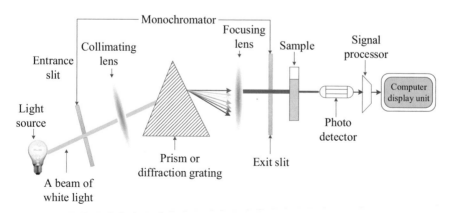

Fig. 3.9 Schematic diagram showing main components of UV-Vis spectrometer

visual/computer display unit. The light produced by the light source is usually a combination of halogen lamps with a wavelength ranging from ~320 to 1100 nm for the visible and near-infrared regions and the deuterium lamps with a wavelength ranging from 190 to 370 nm used for the ultraviolet region.

The monochromator functions as an optical device that selectively transmits a narrow wavelength of light from a broader range of wavelengths available at the input. The monochromator contains other components such as the entrance and exit slits, collimator and focusing lenses and diffraction grating. The entrance slit narrows the white light from the light source incident on the collimator lens, while the collimator lens functions as a device that makes parallel incident light on the diffraction grating. The diffraction grating separates the polychromatic incident light into its component wavelengths (monochromatic light). The monochromatic light is focussed on the sample under investigation through the exit slit which debars the propagation of any stray light on to the sample. Depending on the optical properties of the sample, a fraction of the monochromatic beam intensity is absorbed while the rest is either transmitted or reflected.

The relationship between the incident beam intensity (I_o) on the sample under investigation and the transmitted beam intensity (I) can be represented using Equation 3.10:

$$I(x) = I_o\exp(-\alpha x) \qquad \text{(Equation 3.10)}$$

where x is the thickness of the sample (which is mainly a thin-film semiconductor material in this work) and α is a material constant known as the absorption coefficient which determines the rate of absorption of light by a material as it propagates through it. The transmittance (T) of the sample under investigation can be defined as the ratio of the incident beam intensity (I_o) to the intensity of the transmitted beam (I) (see Equation 3.11):

$$T = \frac{I}{I_o} \qquad \text{(Equation 3.11)}$$

The relationship between the transmittance (T) and absorbance (A) of the sample under investigation as defined in the literature can be represented by Equation 3.12 [32]:

$$A = \log_{10}\left(\frac{I_o}{I}\right) = \log_{10}\left(\frac{1}{T}\right) \qquad \text{(Equation 3.12)}$$

The reflectance R can be obtained using Equation 3.13, where n is the real part of the complex refractive index (N) of semiconductor material:

$$R = \frac{(n-1)^2}{(n+1)^2} \qquad \text{(Equation 3.13)}$$

The complex refractive index N can be obtained using Equation 3.14, where K is the extinction coefficient (the imaginary part of the refractive index):

$$N = n + iK \qquad \text{(Equation 3.14)}$$

As shown in Fig. 3.9, the fraction of the incident monochromatic light transmitted through the sample is detected by the photodetector. The photodetector converts the photons incident on it into current by the generation of electron-hole (e-h) pairs from its p-n junction configuration. For each sampled monochromatic light wavelength, the signal is processed and amplified by the signal processor and displayed on the computer display unit. From the optical property data accumulated using the spectrometer for a specified wavelength range, the bandgap of the semiconductor material can be estimated by plotting a graph of absorbance square (A^2) against photon energy (hv) and extrapolating the straight line portion of the absorption curve to the photon energy axis. Alternatively, Tauc's plot of (αhv)^2 against (hv) can also be utilised [33]. The Tauc's formula is shown in Equation 3.15 which depicts the relationship between absorption coefficient (α), bandgap energy (E_g), Planck's constant (h) and the incident photon frequency (v) for a direct bandgap material, where k is the proportionality constant which depends on the refractive index of the sample under investigation and n is the power factor of the transmission mode. The value m equals 0.5 for direct bandgap and 2.0 for indirect bandgap materials. Based on the literature, sharp absorption edge signifies lesser impurity energy levels and defects in the thin film [34–36]:

$$\alpha = \frac{k\left(hv - E_g\right)^m}{hv} \qquad \text{(Equation 3.15)}$$

With the CdS/CdTe-based solar cells considered in this book, it should be noted that optical losses such as reflection and parasitic absorption have been associated with lowering the conversion efficiency of these solar cells. The predominant loss of ~15% of the attainable short-circuit current density J_{sc} of the fabricated cells is associated with the parasitic absorption of the buffer and/or the window layer [37–39]. Based on literature and other experimental work later discussed in Chap. 6, parasitic absorption can be reduced through buffer and/or window layer thickness optimisation [11, 40, 41]. While reflection losses at the air-glass, glass-transparent conducting oxide (TCO), TCO-CdS and CdS-CdTe account for ~8% loss of the attainable J_{sc} of the fabricated solar cells [37–39]. Reflection losses are reducible to <2% using antireflection coatings [42].

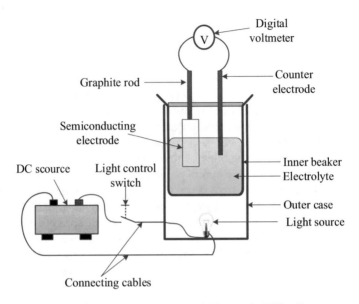

Fig. 3.10 Typical schematic diagram of the experimental setup for PEC cell measurements

3.4.7 Photoelectrochemical (PEC) Cell Characterisation Technique

Photoelectrochemical (PEC) cell measurement is a simple technique utilised for the determination of the conductivity type of a semiconductor material [10]. The preference of PEC cell measurement to the more robust Hall Effect measurement is due to the inclusion of underlying transparent conducting oxide (TCO) substrate on which the semiconductor material is deposited. Due to the high conductivity of the TCO, as compared to the semiconductor layer deposited on it, the TCO serves as an alternative current route under the Hall Effect technique. Therefore the electrical parameters obtained are deemed unreliable due to the influence of the TCO.

Figure 3.10 shows the basic setup of PEC cell measurement which includes DC source, light source, digital voltmeter, semiconductor (sample under investigation), a counter electrode, inner beaker, outer case and electrolyte. The electrolyte utilised for all the PEC measurement presented in this work is 0.10 M $Na_2S_2O_3$ in 20 mL of deionised water. The measurements were taken at ~25 °C.

PEC cell measurement is based on the formation of a semiconductor/electrolyte (S/E) junction similar to an M/S junction when a semiconductor and suitable electrolyte are brought in contact. Due to the difference between the Fermi level E_F of the semiconductor and the energy level of the electrolyte E_{oREDOX}, transfer of electron occurs between the two equilibria positions until a new equilibrium is achieved in which the Fermi level of the semiconductor aligns with the energy level of the electrolyte ($E_{oREDOX} = E_F$). This results in the formation of a Schottky barrier in which the band bending direction and the flow of electron is determined by

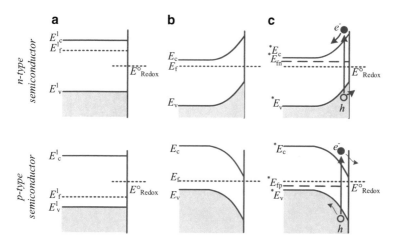

Fig. 3.11 Band diagram formation between semiconductor and electrolyte (**a**) before the semiconductor makes contact with the electrolyte, (**b**) after the semiconductor makes contact with the electrolyte under dark condition and (**c**) after the semiconductor makes contact with the electrolyte under illumination condition

the conductivity type of the semiconductor, as shown in Fig. 3.11. Under the illuminated condition, photons with energy higher than the semiconductor bandgap are absorbed into the depletion region at the S/E junction creating electron-hole (e-h) pairs. The creation of the e-h pairs disrupts the equilibrium position attained ($E_{oREDOX} = E_F$) under dark condition and shifts in the E_F value to a new energy level (E_{Fn} or E_{Fp}) closer to the initial semiconductor Fermi level before contact with the electrolyte [43] as shown in Fig. 3.11c. This results in reductions in band bending and potential change due to Fermi level position alteration (ΔE) which corresponds to Equation 3.16 [43] for measurements taken under illuminated and dark conditions:

$$\Delta E = \left(\frac{1}{e}\right) \left| *EF - E_F \right| = E_{\text{light}} - E_{\text{dark}} \qquad \text{(Equation 3.16)}$$

Based on this understanding, the voltage across the semiconductor and the counter electrode are recorded under both dark and illuminated conditions. The difference between the voltages measured under illuminated (V_L) and dark (V_D) conditions represents the PEC signal or open-circuit voltage of the semiconductor/electrolyte junction. It should be noted that the PEC signal may give a zero value due to wide bandgap range in insulators, overlapping bandgap in conductors (metal) or the mid-gap positioning of the Fermi level in an intrinsic semiconductor.

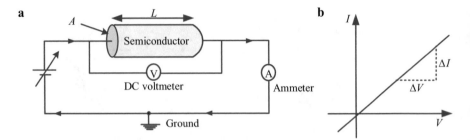

Fig. 3.12 (**a**) Typical schematic circuit diagram illustrating the direction of current flow in semiconductor as the voltage is being varied and, (**b**) shows the typical *I-V* characteristic of a semiconductor material that obeys the Ohmic law. Note the linear relationship between the direct current and the voltage

3.4.8 Direct Current Conductivity Measurement Technique

The DC conductivity measurement utilises the Ohm's law concept in determining the electrical resistivity (ρ) and conductivity (σ) of semiconductor materials. This is achieved by forming two ohmic contacts with the semiconductor material through careful metal contact selection. For ohmic contacts to be formed, the difference between metal work function ϕ_m and semiconductor electron affinity χ_s should not exceed ~0.40 eV [10]. In this technique, varying DC voltages between -1.00 and 1.00 V are applied at constant steps across the two semiconductor terminals. The corresponding DC current that flows through the semiconductor material is measured using an ammeter. In this research work, the DC conductivity measurements were carried out using Keithley 2401 SourceMeter.

The obtained (*I-V*) data give a straight line graph passing through the origin (0, 0) as shown in Fig. 3.12. The slope inverse of the straight line graph gives the resistance R as shown in Equation 3.17:

$$R = \frac{\Delta V}{\Delta I} \qquad \text{(Equation 3.17)}$$

It is well known that Resistance R and Resistivity ρ can be related according to Equation 3.18 [44]:

$$R = \frac{\rho L}{A} \qquad \text{(Equation 3.18)}$$

which can be rearranged as Equation 3.19:

$$\rho = \frac{RA}{L} \qquad \text{(Equation 3.19)}$$

where ρ (Ω cm) is the electrical resistivity, A (cm^2) is the cross-sectional area and L (cm) is the thickness. Furthermore, resistivity ρ (Ω cm) and conductivity σ (Ω cm)$^{-1}$ are related according to Equation 3.20:

$$\sigma = \frac{1}{\rho} \qquad\qquad \text{(Equation 3.20)}$$

3.5 Device Characterisation Techniques

After the optimisation of the semiconductor layers based on their material properties, electronic devices such as solar cells were fabricated. The electronic properties of the fabricated solar cell devices were explored using both the current-voltage (*I-V*) and capacitance-voltage (*C-V*) techniques. For both techniques, the bias range is set between −1.00 and +1.00 V, and the resulting current and capacitance are measured, respectively. Both the *I-V* and *C-V* tests are non-destructive tests. The *I-V* characteristics were explored under both dark and illuminated (AM1.5) conditions, while the *C-V* characteristics were explored under dark condition. As noted above, due to the underlying transparent conducting oxide (TCO) which is a prerequisite for electrodeposition technique, robust electronic characterisation technique such as Hall Effect measurement is unsuitable for the acquisition of the electrical parameters required. Hence, the *I-V* and *C-V* characterisation techniques were explored as an alternative method to extract material properties such as doping concentration and charge carrier mobility.

3.5.1 Current-Voltage (I-V) Characterisation

The use of current-voltage (*I-V*) technique for Schottky diode characterisation has been well explored in the determination of semiconductor electronic properties under both dark and AM1.5 illuminated conditions. Under dark condition, the obtained *I-V* data can be plotted as log-linear or linear-linear graphs. From the log-linear graphs, device parameters such as rectification factor (RF), ideality factor (n), saturation current (I_o) and potential barrier height (ϕ_b) can be determined, while shunt (R_{sh}) and series (R_s) resistances can be determined from linear-linear *I-V* graphs. Other solar cell parameters such as open-circuit voltage (V_{oc}), short-circuit current density (J_{sc}), fill factor (FF) and conversion efficiency (η) can be determined from the linear-linear *I-V* curve measured under AM1.5 illuminated condition. It should be noted that R_{sh} and R_s can be evaluated under both dark and AM1.5 conditions, but due to the possible fluctuations of irradiation intensity under illuminated condition and its effect on other diode parameters [45], the R_{sh} and R_s obtained under dark condition is considered to be more viable.

A typical linear-linear *I-V* curve under illuminated condition iterating the effect of R_{sh} and R_s on the *I-V* curve and their equivalent circuits are shown in Fig. 3.13. For an ideal diode, as shown in Fig. 3.13 scenario (a), the R_{sh} and R_s equal to infinity and zero, respectively [44].

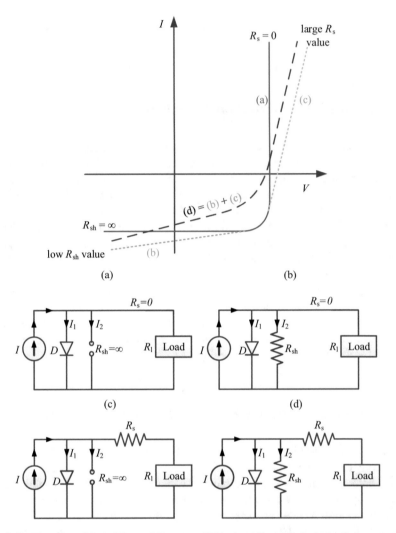

Fig. 3.13 The effect of R_s and R_{sh} on I-V curves of PV solar cells and their single-diode equivalent circuits

In this case, the R_{sh} behaves like an open-circuit voltage, and all the photo-generated currents I flow only through the load (represented by diode D) and result in the maximum possible current output. It should be noted that reductions in the R_{sh} as shown in Fig. 3.13 scenario (b) will lead to the creation of an alternative flow path for current through which a fraction of the photo-generated current is lost through shunt. By Kirchhoff's law, the total photo-generated current I divide into I_1 and I_2. I_1 goes through the solar cell represented by a diode, D, while I_2 goes through the shunt path. Reductions in R_{sh} which is often related to the semiconductor material quality may also result in the reduction of other solar cell parameters such as FF and V_{oc} [46].

As shown in Fig. 3.13 scenario (c), an increase in R_s results in voltage drop before the load. R_s is associated with M/S contact resistance, usage of highly resistive semiconductor material and the presence of excess oxide layer in between the M/S interface [47]. The main impact of R_s is the reduction of photon to electron conversion efficiency via J_{sc} and FF reduction [47]. The combination of scenarios (b) and (c) as shown in Fig. 3.13 scenario (d) with low R_{sh} and high R_s results in a gradual reduction in V_{oc}, J_{sc} and FF which ultimately affects the efficiency of the solar cell.

3.5.1.1   I-V Characterisation Under Dark Condition

Under dark conditions, solar cell devices behave like diodes due to the formation of a semiconductor p-n junction or a M/S Schottky contact. Typical linear-linear and log-linear I-V characteristic curves are shown in Fig. 3.14. The I-V characteristics of a Schottky-type diode under dark condition can be expressed by Equation 3.21 [48]:

$$I_D = SA^*T^2 \cdot \exp\left(\frac{-e\phi_b}{kT}\right)\left[\exp\left(\frac{eV}{nkT}\right) - 1\right] \qquad \text{(Equation 3.21)}$$

where I_D is the electric current in dark condition (A), I_0 is the reverse saturation current (A), S is the area of the contact (cm^2), A^* is the effective Richardson constant for thermionic emission (Acm^{-2} K^{-2}), T is the temperature (K), e is the electronic charge (1.60×10^{-19} C), ϕ_b is the potential barrier height (eV), k is the Boltzmann constant (1.38×10^{-23} JK^{-1}) and n is the ideality factor of the diode:

$$\text{since} \quad I_0 = SA^*T^2 \cdot \exp\left(\frac{-e\phi_b}{kT}\right) \qquad \text{(Equation 3.22)}$$

$$\text{or} \quad \phi_b = \frac{kT}{e}\ln\left(\frac{SA^*T^2}{I_0}\right) \qquad \text{(Equation 3.23)}$$

Equation 3.21 can be rewritten as:

$$I_D = I_0\left[\exp\left(\frac{eV}{nkT}\right) - 1\right] \qquad \text{(Equation 3.24)}$$

As reported by Rhoderick and Williams 1982 [48], if the applied voltage across the diode exceeds 75 mV, then:

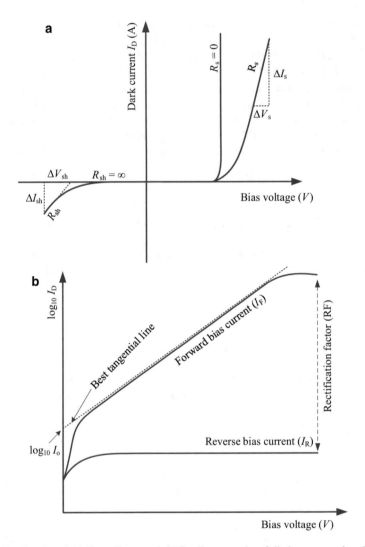

Fig. 3.14 The plot of (**a**) linear-linear and (**b**) log-linear graphs of diodes measured under dark condition

$$\exp\left(\frac{eV}{nkT}\right) \gg 1 \qquad \text{(Equation 3.25)}$$

Equation 3.24 can be rewritten as:

$$I_D = I_0 \cdot \exp\left(\frac{eV}{nkT}\right) \qquad \text{(Equation 3.26)}$$

$$\text{or} \quad \log_{10}(I_D) = \left(\frac{e}{2.303\,nkT}\right) \cdot V + \log_{10}(I_0) \qquad \text{(Equation 3.27)}$$

$$\text{or} \quad \ln I_D = \left(\frac{e}{nkT}\right)V + \ln I_0 \qquad \text{(Equation 3.28)}$$

Therefore, from the plot of $\log_{10}(I_D)$ versus voltage applied (V) as shown in Fig. 3.14b, parameters such as I_0, rectification factor (RF), ϕ_b and n can be obtained. The intercept of the best forward-bias tangential line on the current axis gives the saturation current I_0, while the slope in the forward bias gives the value through which the ideality factor n was calculated using Equation 3.30 with the consideration that the largest gradient of the curve gives the lowest value of n [48]. The n value provides information on the current transport mechanisms such that for an ideal diode, the current transport mechanism is through thermionic emission ($n = 1.00$). When the value of n is 2.00, then the current transport mechanism is entirely through recombination and generation (R&G), but if n is between 1.00 and 2.00, then the current transport mechanism consists of both thermionic emission and R&G [48]. For n values above 2.00, the current transportation mechanism is not limited to thermionic emission and R&G but also due to the tunnelling of high-energy electron through the barrier height [49]. Large R_s value of the diode also contributes to the increase in n values:

$$\text{slope} = \frac{\Delta \log_{10} I}{\Delta V} = \frac{e}{2.303\,nkT} \qquad \text{(Equation 3.29)}$$

$$n = \frac{e}{kT}\left(\frac{1}{2.303 \times \text{slope}}\right) \simeq \frac{16.78}{\text{slope}} \qquad \text{(Equation 3.30)}$$

The rectification factor RF which helps in determining the quality of a rectifying diode can be calculated as the ratio of forward current to reverse current at a constant voltage as estimated from the $\log_{10}(I_D)$ versus voltage (V) plot. As documented in the literature, RF values at an excess of 10^3 are sufficient for a high-performance solar cell [10]. The barrier height ϕ_b was calculated using Equation 3.23, and the effective Richardson constant A^* was calculated using Equation 3.31:

$$A^* = \frac{4\pi m^* k^2 q}{h^3} \qquad \text{(Equation 3.31)}$$

where m^* is the effective electron mass and h is the Planck's constant (6.626×10^{-30} cm^{2}kgs^{-1}). Depending on the conductivity type of the semiconductor, the effective electron mass can be denoted as m_e^* or m_p^* for n-type and p-type semiconductor material, respectively. The effective electron masses of n-CdS, n-CdTe and p-CdTe are given by $0.21m_o$, $0.1m_o$ and $0.63m_o$ [50, 51], respectively. The free electron mass $m_o = 9.1 \times 10^{-31}$ kg.

3.5.1.2 *I-V* Characterisation Under Illumination

Under illuminated conditions, the direction of flow of photo-generated current I_{SC} is opposite to the forward current due to bias voltage under dark condition, I_D. Therefore, the total current of a solar cell under illumination is given by Equation 3.32:

$$I_L = I_D - I_{SC} \qquad \text{(Equation 3.32)}$$

With the incorporation of I_D from Equation 3.24, Equation 3.32 can be rewritten as:

$$I_L = I_0 \left[\exp\left(\frac{eV}{nkT}\right) - 1 \right] - I_{sc} \qquad \text{(Equation 3.33)}$$

Under open-circuit condition, when $I_L = 0$ and $V = V_{oc}$ as shown in Fig. 3.15, Equation 3.33 can be rewritten as:

$$0 = I_0 \left[\exp\left(\frac{eV_{oc}}{nkT}\right) - 1 \right] - I_{sc} \qquad \text{(Equation 3.34)}$$

$$\text{or} \quad I_{sc} = SA^*T^2 \left[\exp\left(\frac{-e\phi_b}{kT}\right) \right] \left[\exp\left(\frac{eV_{oc}}{nkT}\right) - 1 \right] \qquad \text{(Equation 3.35)}$$

By taking the natural logarithm of Equation 3.35 and noting that:

Fig. 3.15 Typical *I-V* curve of solar cells measured under dark and illuminated conditions

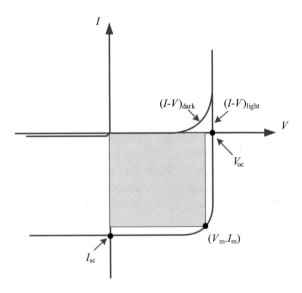

$$J_{sc} = \frac{I_{sc}}{S}$$ (Equation 3.36)

Equation 3.35 can be rewritten as:

$$V_{oc} = n\left[\phi_b + \frac{kT}{e}\ln\left(\frac{J_{sc}}{A^*T^2}\right)\right]$$ (Equation 3.37)

Equation 3.37 shows the dependence of increasing V_{oc} value on the increase of n and ϕ_b values. However, due to the significance of n value on current transport mechanisms and its effect on J_{sc}, a limiting value of $n \leq 2.00$ is desirable for high-efficiency solar cell devices. It should be noted that in Schottky diodes, the V_{oc} of a solar cell device can also be improved by incorporating thin insulating layer between metal-semiconductor (M/S) contact to give an MIS structure [52, 53] as discussed in Sect. 2.3.5, in which both the V_{oc} and the ϕ_b show considerable improvements. Furthermore, reduction in the temperature of solar cells can also improve V_{oc} value as a reduction in temperature minimises thermal agitation of the charge carriers [10, 54].

Another parameter which can be determined from the I-V curve under AM1.5 condition as shown in Fig. 3.15 is the fill factor FF. The FF can be defined as the ratio of the maximum power P_{max} of the solar cell to the product of I_{sc} and V_{oc} as shown in Equation 3.38. Improvements in FF can be achieved with an increase in R_{sh} and reduction in R_s values. Maximum FF value can be attained under ideal diode condition where R_{sh} is ∞ and $R_s = 0\ \Omega$:

$$FF = \frac{P_{max}}{I_{sc}V_{oc}} = \frac{I_m V_m}{I_{sc}V_{oc}}$$ (Equation 3.38)

The efficiency of the solar cell (η) which can be defined as the ratio of power output P_{output} to power input P_{input} can be illustrated as shown in Equation 3.39:

$$\eta = \frac{\text{Power output}}{\text{Power input}} = \frac{V_m I_m}{P_{in}} = \frac{V_{oc} \cdot I_{sc} \cdot FF}{P_{in}}$$ (Equation 3.39)

Under standard AM1.5 illuminated condition, P_{input} is $1000\ \text{Wm}^{-2}$ ($100\ \text{mWcm}^{-2}$), therefore, Equation 3.39 can be rewritten as:

$$\eta = \frac{V_{oc} \cdot J_{sc} \cdot FF}{P_{in}}$$ (Equation 3.40)

All the I-V measurements for both AM1.5 illuminated ($1000\ \text{Wm}^{-2}$) and dark conditions as reported in this book were performed using a Rera Solution I-V measurement system in the ambient environment. The Rera Solution I-V measurement system is constituted of Keithley 2401 SourceMeter, LOT Quantum Design GmbH arc light source and a computer laptop running Rera Tracer IV software. The I-V system was precalibrated using RR267MON standard silicon solar cell.

3.5.2   Capacitance-Voltage (C-V) Characterisation

Capacitance-Voltage (C-V) technique has been well explored in the literature and utilised for the determination of diode electronic properties, in which parameters such as capacitance at zero bias (C_o) can be directly determined from the C-V curve while other parameters such as Fermi level (E_F) position with respect to both the conduction (E_C) and valence bands (E_V), built-in potential (V_{bi}), barrier height ϕ_b, doping concentration of donors (N_D) or acceptors (N_A) and the depletion width at zero bias (W) [44] can be determined using Mott-Schottky plots. The Mott-Schottky plot is a graph of C^{-2} against V. For this report, the C-V technique was performed at 1 MHz due to the reduction in defect interference at high frequency [55] and under dark conditions due to noise caused by the high capture and emission of electrons at R&G centres [56] at room temperature. By combining the doping concentration with the DC conductivity, the charge carrier mobility ($\mu_\perp$) can be determined.

Schottky or p-n or junctions are identical to parallel plate capacitors. In which, the junction contains immobile ionised dopants within the depletion region and the semiconductor material (in the case of a p-n junction) serves as the electrode or conducting plates of the capacitor. The charged layers generated at the junction due to potential barrier formation are known as the junction capacitor. The junction capacitance C of the depletion region is given by Equation 3.41:

$$C = \frac{\varepsilon_s A}{W} = \frac{\varepsilon_o \varepsilon_r A}{W}$$
(Equation 3.41)

where $\varepsilon_s = \varepsilon_o \varepsilon_r$ is the permittivity of semiconductor (Fcm^{-1}), ε_o is the permittivity of free space (8.85×10^{-14} Fcm^{-1}), ε_r is the relative permittivity (or the dielectric constant) of the semiconductor material (ε_r is 8.9 for ZnS, 8.9 for CdS and 11.0 for CdTe [57]), A is the junction contact area and W is the width of depletion region (cm). The capacitance can also be expressed per unit area (C_A). Therefore Equation 3.41 can be rewritten as Equation 3.42:

$$C_A = \frac{C}{A} = \frac{\varepsilon_s}{W} = \frac{\varepsilon_o \varepsilon_r}{W}$$
(Equation 3.42)

The built-in potential can be estimated using Equation 3.43 and graphically shown in Fig. 3.17 [58]:

$$V_{bi} = \frac{kT}{e} \ln \left(\frac{N_A N_D}{n_i^2} \right)$$
(Equation 3.43)

and the depletion width W is given by:

$$W = \sqrt{\frac{2\varepsilon_s}{e}\left(\frac{1}{N_A} + \frac{1}{N_D}\right)V_{bi}} \qquad \text{(Equation 3.44)}$$

where N_A is the acceptor concentration, N_D is the donor concentration and n_i is the intrinsic concentration. The intrinsic concentration n_i defines the steady-state situation of an intrinsic semiconductor under thermal agitation and constant electron excitation from E_v to E_c, in which the number of electrons in the conduction band equals the number of holes in the valence band.

For one-sided abrupt junction where the p-type semiconductor material is highly doped ($N_A \gg N_D$), the depletion region extends more into the n-type semiconductor material. Conversely, if the n-type semiconductor material is highly doped ($N_D \gg N_A$), the depletion region extends more into the p-type semiconductor material. Therefore, Equation 3.44 can be rewritten as Equation 3.45, where $N = N_D$ for $N_A \gg N_D$ and $N = N_A$ for $N_D \gg N_A$:

$$W = \sqrt{\frac{2\varepsilon_s V_{bi}}{eN}} \qquad \text{(Equation 3.45)}$$

It should be noted that Equations 3.44 and 3.45 hold when there is no externally applied voltage (V) across the junction. In the presence of an externally applied voltage across the junction, the total voltage across the junction is ($V_{bi} - V$), where, V is positive for forward bias and negative for reverse bias [44]. Therefore, the depletion width (W) under externally applied voltage for the one-sided abrupt junction is given by Equation 3.46:

$$W = \sqrt{\frac{2\varepsilon_s(V_{bi} - V)}{eN}} \qquad \text{(Equation 3.46)}$$

Consequently, Equation 3.41 can be rewritten as Equation 3.47:

$$C = \frac{\varepsilon_s A}{\sqrt{\frac{2\varepsilon_s(V_{bi}-V)}{eN}}} = A\sqrt{\frac{e\varepsilon_s N}{2(V_{bi} - V)}} \qquad \text{(Equation 3.47)}$$

With respect to the depletion capacitance per unit area as shown in Equation 3.42, Equation 3.47 can be redefined as Equation 3.48.

$$C_A = \frac{C}{A} = \sqrt{\frac{e\varepsilon_s N}{2(V_{bi} - V)}} \qquad \text{(Equation 3.48)}$$

Figure 3.16 shows the typical graph of capacitance per unit area (C_A) against bias voltage (V) from which the C_o capacitance per unit area at zero bias can be determined. The obtained C_o value from this graph is utilised in the calculation of the depletion width at zero bias using Equation 3.41.

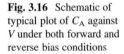

Fig. 3.16 Schematic of typical plot of C_A against V under both forward and reverse bias conditions

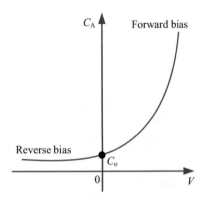

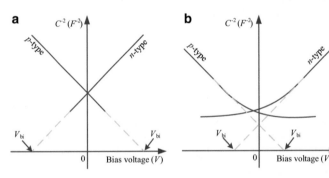

Fig. 3.17 Typical Mott-Schottky plots of p- and n-type semiconductors (**a**) for an ideal diode and (**b**) for a non-ideal diode

The equation defining the Mott-Schottky plot from which other parameters are determined can be achieved by squaring both sides of Equation 3.48 and rearranging it to give Equation 3.49, where V is either positive or negative for forward or reverse bias, respectively:

$$C_A^2 = \frac{e\varepsilon_s N}{2(V_{bi} - V)} \quad \text{or} \quad C^2 = \frac{e\varepsilon_s NA^2}{2(V_{bi} - V)} \qquad \text{(Equation 3.49)}$$

Equation 3.49 can also be expressed as Equation 3.50:

$$\frac{1}{C^2} = C^{-2} = \frac{2}{e\varepsilon_s NA^2}(V_{bi} - V) \qquad \text{(Equation 3.50)}$$

As observed in Equation 3.50, a plot of C^{-2} against V for one-sided abrupt junction should produce a straight line [44] as shown in Fig. 3.17a due to its equivalence to straight line equation ($y = mx + c$) where m is the slope. But as a result of the effects of defects, traps, surface states and interfacial resistive layers (attributable to oxidation) producing excess capacitance and inhomogeneity of the semiconductor layer in the diode, deviation from linearity is usually observed [59] as shown in Fig. 3.17b. The slope of the C^{-2} against V plot can be defined as shown in Equation 3.51, and the extrapolation to $C^{-2} = 0$ gives the built-in potential (V_{bi}).

$$slope = \frac{2}{e\varepsilon_s NA^2} \qquad \text{(Equation 3.51)}$$

Therefore, with the known value m (slope) and V_{bi} (intercept) from the Mott-Schottky plot, the doping density (N) can be calculated by rearranging Equation 3.51 as Equation 3.52.

$$N = \frac{2}{\varepsilon_s e A^2 \cdot slope} \qquad \text{(Equation 3.52)}$$

The effective density of states both in the conduction (N_c) and valence (N_v) bands are respectively shown in Equations 3.53 and 3.54. The equations show the dependence of the effective density of states on temperature and the nature of semiconductor material. Therefore, the N_c and N_v values differ for different semiconductor materials and conductivity types:

$$N_c = 2\left(\frac{2\pi m_e^* kT}{h^2}\right)^{3/2} \qquad \text{(Equation 3.53)}$$

$$N_v = 2\left(\frac{2\pi m_p^* kT}{h^2}\right)^{3/2} \qquad \text{(Equation 3.54)}$$

where m_e^* and m_p^* are the electron and hole effective masses for n-type and p-type semiconductor materials, respectively (with all parameters predefined in Sect. 3.5.1.1).

With the doping density N_A or N_D (Equation 3.52) and the effective density of states N_c and N_v as calculated from Equation 3.53 or Equation 3.54 derived from Mott-Schottky plot, the Fermi level E_F, the barrier height ϕ_b and the charge carrier mobility $\mu_\perp$ can be calculated with the assumption that all excess donor (N_D) or acceptor (N_A) atoms are completely ionised at room temperature as shown in Equations 3.55–3.58. Therefore, the Fermi Dirac probability function of electrons occupying the donor state is $n \approx (N_D - N_A)$. The electrical conductivity (σ) values from the DC conductivity measurement were used to estimate $\mu_\perp$ for the semiconductor. It should be noted that the $\mu_\perp$ value is the mobility of charge carriers normal to the surface of the thin films:

$$\Delta E = E_C - E_F = \frac{kT}{e} \ln\left(\frac{N_C}{N_D - N_A}\right) \qquad \text{(Equation 3.55)}$$

$$\text{or} \quad \Delta E = E_F - E_V = \frac{kT}{e} \ln\left(\frac{N_V}{N_A - N_D}\right) \qquad \text{(Equation 3.56)}$$

$$\phi_{\mathrm{b}} = V_{\mathrm{bi}} + E_{\mathrm{F}} + \frac{kT}{e} \qquad \text{(Equation 3.57)}$$

$$\mu_{\perp} = \frac{\sigma}{ne} = \frac{\sigma}{(N_{\mathrm{D}} - N_{\mathrm{A}})e} \qquad \text{(Equation 3.58)}$$

The *C-V* measurements as reported in this book were performed using Hewlett Packard 4284A 20 Hz–1 MHz Precision LCR Meter (Yokogawa Hewlett Packard, Japan) with a Keithley 6517A Electrometer/High Resistance Meter.

3.6 Conclusion

This chapter discussed the main semiconductor deposition technique utilised in this work. It further presented the techniques employed in the analysis and investigation of the deposition voltage, elemental deposition sequence, structural, morphological, compositional, optical and the electrical properties of the semiconductors. The analysis and investigation are aimed at optimising both growth and treatment conditions for device purposes. Further to material properties, both *I-V* and *C-V* techniques were also utilised for the characterisation of the fabricated devices. The results presented in this book using the techniques mentioned above were repeatedly performed to ascertain the reproducibility of similar trends.

References

1. K.L. Chopra, P.D. Paulson, V. Dutta, Thin-film solar cells: an overview. Prog. Photovolt. Res. Appl. **12**, 69–92 (2004). https://doi.org/10.1002/pip.541
2. R.A. Street, Thin-film transistors. Adv. Mater. **21**, 2007–2022 (2009). https://doi.org/10.1002/adma.200803211
3. W. Kern, K.K. Schuegraf, in *Handb. Thin Film Depos. Process. Tech.* Deposition technologies and applications (Elsevier, New York, 2001), pp. 11–43. https://doi.org/10.1016/B978-081551442-8.50006-7
4. D. Lincot, Electrodeposition of semiconductors. Thin Solid Films **487**, 40–48 (2005). https://doi.org/10.1016/j.tsf.2005.01.032
5. A.A. Ojo, I.M. Dharmadasa, The effect of fluorine doping on the characteristic behaviour of CdTe. J. Electron. Mater. **45**, 5728–5738 (2016). https://doi.org/10.1007/s11664-016-4786-9
6. J.M. Woodcock, A.K. Turner, M.E. Ozsan, J.G. Summers, in *Conf. Rec. Twenty-Second IEEE Photovolt. Spec. Conf.—1991*. Thin film solar cells based on electrodeposited CdTe (IEEE, 1991), pp. 842–847. https://doi.org/10.1109/PVSC.1991.169328
7. D. Cunningham, M. Rubcich, D. Skinner, Cadmium telluride PV module manufacturing at BP Solar. Prog. Photovolt. Res. Appl. **10**, 159–168 (2002). https://doi.org/10.1002/pip.417
8. I.M. Dharmadasa, J. Haigh, Strengths and advantages of electrodeposition as a semiconductor growth technique for applications in macroelectronic devices. J. Electrochem. Soc. **153**, G47 (2006). https://doi.org/10.1149/1.2128120

9. A.A. Ojo, I.M. Dharmadasa, Progress in development of graded bandgap thin film solar cells with electroplated materials. J. Mater. Sci. Mater. Electron. **28**, 6359–6365 (2017). https://doi.org/10.1007/s10854-017-6366-z

10. I.M. Dharmadasa, *Advances in Thin-Film Solar Cells* (Pan Stanford, Singapore, 2013)

11. A.A. Ojo, H.I. Salim, O.I. Olusola, M.L. Madugu, I.M. Dharmadasa, Effect of thickness: a case study of electrodeposited CdS in CdS/CdTe based photovoltaic devices. J. Mater. Sci. Mater. Electron. **28**, 3254–3263 (2017). https://doi.org/10.1007/s10854-016-5916-0

12. M.P.R. Panicker, M. Knaster, F.A. Kroger, Cathodic deposition of CdTe from aqueous electrolytes. J. Electrochem. Soc. **125**, 566 (1978). https://doi.org/10.1149/1.2131499

13. J. McHardy, F. Ludwig, *Electrochemistry of semiconductors and electronics: processes and devices* (Noyes Publications, Park Ridge, 1992). https://books.google.co.uk/books?id=cSEt5W3vmdIC

14. A.A. Ojo, I.M. Dharmadasa, 15.3% Efficient graded bandgap solar cells fabricated using electroplated CdS and CdTe thin films. Sol. Energy **136**, 10–14 (2016). https://doi.org/10.1016/j.solener.2016.06.067

15. I.M. Dharmadasa, P. Bingham, O.K. Echendu, H.I. Salim, T. Druffel, R. Dharmadasa, G. Sumanasekera, R. Dharmasena, M.B. Dergacheva, K. Mit, K. Urazov, L. Bowen, M. Walls, A. Abbas, Fabrication of CdS/CdTe-based thin film solar cells using an electrochemical technique. Coatings. **4**, 380–415 (2014). https://doi.org/10.3390/coatings4030380

16. J. Pandey, Solar cell harvesting: green renewable technology of future introduction. Int. J. Adv. Res. Eng. Appl. Sci. **4**, 93 (2015). ISSN: 2278–6252

17. S. Dennison, Dopant and impurity effects in electrodeposited CdS/CdTe thin films for photovoltaic applications. J. Mater. Chem. **4**, 41 (1994). https://doi.org/10.1039/jm9940400041

18. K. Zanio, *Semiconductors and Semimetals* (Academic Press, New York, 1978). http://shu.summon.serialssolutions.com/2.0.0/link/0/eLvHCXMwdV3JCsIwEB1cEAQPrrgV-gNKmyZNPYvFu94l6bQ3K1j_HydDXXA5Zg7DJJB5me0FIBLrYPXhE8LEUZ8lRggjsgBlgBuptSqw0Chzrsy80Rg848ZXCuObQZ_iCKmCyN3HJjQJON2LqOaiYzdM7pnQiml0hCaUCiNVM-481sn7lwYMKGkfWm7IYACNvBxCh9sws2o

19. E.A. Meulenkamp, L.M. Peter, Mechanistic aspects of the electrodeposition of stoichiometric CdTe on semiconductor substrates. J. Chem. Soc. Trans. **92**, 4077–4082 (1996). https://doi.org/10.1039/ft9969204077

20. A.Y. Shenouda, E.S.M. El Sayed, Electrodeposition, characterization and photo electrochemical properties of CdSe and CdTe. Ain Shams Eng. J. **6**, 341–346 (2014). https://doi.org/10.1016/j.asej.2014.07.010

21. H.Y.R. Atapattu, D.S.M. De Silva, K.A.S. Pathiratne, I.M. Dharmadasa, Effect of stirring rate of electrolyte on properties of electrodeposited CdS layers. J. Mater. Sci. Mater. Electron. **27**, 5415–5421 (2016). https://doi.org/10.1007/s10854-016-4443-3

22. P.T. Kissinger, W.R. Heineman, Cyclic voltammetry. J. Chem. Educ. **60**, 702 (1983). https://doi.org/10.1021/ed060p702

23. C.W. Siders, Detection of nonthermal melting by ultrafast X-ray diffraction. Science **286**, 1340–1342 (1999). https://doi.org/10.1126/science.286.5443.1340

24. A. Monshi, Modified Scherrer equation to estimate more accurately nano-crystallite size using XRD. World J. Nano Sci. Eng. **2**, 154–160 (2012). https://doi.org/10.4236/wjnse.2012.23020.

25. H.J. Butler, L. Ashton, B. Bird, G. Cinque, K. Curtis, J. Dorney, K. Esmonde-White, N.J. Fullwood, B. Gardner, P.L. Martin-Hirsch, M.J. Walsh, M.R. Mcainsh, N. Stone, F.L. Martin, Using Raman spectroscopy to characterise biological materials. Nat. Protoc. **11**, 664–687 (2016). https://doi.org/10.1038/nprot.2016.036

26. A.K. Yadav, P. Singh, A review of the structures of oxide glasses by Raman spectroscopy. RSC Adv. **5**, 67583–67609 (2015). https://doi.org/10.1039/C5RA13043C

27. W.A. Mackie, G.G. Magera, Defined emission area and custom thermal electron sources. J. Vac. Sci. Technol. B, Nanotechnol. Microelectron. Mater. Process. Meas. Phenom. **29**, 06F601 (2011). https://doi.org/10.1116/1.3656350

28. D.N. Leonard, G.W. Chandler, S. Seraphin, in *Charact. Mater.*. Scanning electron microscopy (Wiley, Hoboken, 2012), pp. 1721–1735. https://doi.org/10.1002/0471266965.com081.pub2.

29. D.A. Moncrieff, P.R. Barker, Secondary electron emission in the scanning electron microscope. Scanning **1**, 195–197 (1978). https://doi.org/10.1002/sca.4950010307

30. K. Kanaya, S. Okayama, Penetration and energy-loss theory of electrons in solid targets. J. Phys. D. Appl. Phys. **5**, 308 (1972). https://doi.org/10.1088/0022-3727/5/1/308

31. P.J. Statham, Limitations to accuracy in extracting characteristic line intensities from x-ray spectra. J. Res. Natl. Inst. Stand. Technol. **107**, 531 (2002). https://doi.org/10.6028/jres.107.045

32. H.C. Allen, T. Brauers, B.J. Finlayson-Pitts, Illustration of deviations in the Beer-Lambert law in an instrumental analysis laboratory: measuring atmospheric pollutants by differential optical absorption spectrometry. J. Chem. Educ. **74**, 1459 (1997). https://doi.org/10.1021/ed074p1459

33. J. Tauc, Optical properties and electronic structure of amorphous Ge and Si. Mater. Res. Bull. **3**, 37–46 (1968). https://doi.org/10.1016/0025-5408(68)90023-8

34. J. Han, C. Spanheimer, G. Haindl, G. Fu, V. Krishnakumar, J. Schaffner, C. Fan, K. Zhao, A. Klein, W. Jaegermann, Optimized chemical bath deposited CdS layers for the improvement of CdTe solar cells. Sol. Energy Mater. Sol. Cells **95**, 816–820 (2011). https://doi.org/10.1016/j.solmat.2010.10.027.

35. A. Bosio, N. Romeo, S. Mazzamuto, V. Canevari, Polycrystalline CdTe thin films for photovoltaic applications. Prog. Cryst. Growth Charact. Mater. **52**, 247–279 (2006). https://doi.org/10.1016/j.pcrysgrow.2006.09.001

36. D.T.F. Marple, Optical absorption edge in CdTe: experimental. Phys. Rev. **150**, 728–734 (1966). https://doi.org/10.1103/PhysRev.150.728

37. V.Y. Roshko, L. a Kosyachenko, E.V. Grushko, Theoretical analysis of optical losses in CdS/CdTe solar cells. Acta Phys. Pol. A. **120**, 954–956 (2011). http://przyrbwn.icm.edu.pl/APP/PDF/120/a120z5p39.pdf

38. H.A. Mohamed, Dependence of efficiency of thin-film CdS/CdTe solar cell on optical and recombination losses. J. Appl. Phys. **113**, 093105 (2013). https://doi.org/10.1063/1.4794201

39. H.A. Mohamed, Influence of the optical and recombination losses on the efficiency of CdS/CdTe solar cell at ultrathin absorber layer. Can. J. Phys. **92**, 1350–1355 (2014). https://doi.org/10.1139/cjp-2013-0477

40. J.E. Granata, J.R. Sites, in *Conf. Rec. Twenty Fifth IEEE Photovolt. Spec. Conf.*. Effect of CdS thickness on CdS/CdTe quantum efficiency, vol 2000 (1996), pp. 853–856. https://doi.org/10.1109/PVSC.1996.564262

41. J.S. Lee, Y.K. Jun, H.B. Im, Effects of CdS film thickness on the photovoltaic properties of sintered CdS / CdTe solar cells. J. Electrochem. Soc. **134**, 248–251 (1987). https://doi.org/10.1149/1.2100417.

42. P.M. Kaminski, F. Lisco, J.M. Walls, Multilayer broadband antireflective coatings for more efficient thin film CdTe solar cells. IEEE J. Photovolt. **4**, 452–456 (2014). https://doi.org/10.1109/JPHOTOV.2013.2284064

43. K. Rajeshwar, in *Encycl. Electrochem.*. Fundamentals of semiconductor electrochemistry and photoelectrochemistry (Wiley-VCH Verlag GmbH & Co. KGaA, Weinheim, 2007), pp. 1–51. https://doi.org/10.1002/9783527610426.bard060001

44. S.M. Sze, K.K. Ng, *Physics of Semiconductor Devices* (Wiley, Hoboken, 2006). https://doi.org/10.1002/0470068329

45. M. Chegaar, A. Hamzaoui, A. Namoda, P. Petit, M. Aillerie, A. Herguth, Effect of illumination intensity on solar cells parameters. Energy Procedia **36**, 722–729 (2013). https://doi.org/10.1016/j.egypro.2013.07.084

46. T. Soga, *Nanostructured materials for solar energy conversion*, vol 614 (Elsvier Science, Philadelphia, 2006). https://www.elsevier.com/books/nanostructured-materials-for-solar-energy-conversion/soga/978-0-444-52844-5

47. M. Dadu, A. Kapoor, K.N. Tripathi, Effect of operating current dependent series resistance on the fill factor of a solar cell. Sol. Energy Mater. Sol. Cells **71**, 213–218 (2002). https://doi.org/10.1016/S0927-0248(01)00059-9

48. E.H. Rhoderick, Metal-semiconductor contacts. IEE Proc. I Solid State Electron Devices. **129**, 1 (1982). https://doi.org/10.1049/ip-i-1.1982.0001

49. J. Verschraegen, M. Burgelman, J. Penndorf, Temperature dependence of the diode ideality factor in CuInS2-on-Cu-tape solar cells. Thin Solid Films **480–481**, 307–311 (2005). https://doi.org/10.1016/j.tsf.2004.11.006

50. S. Geyer, V.J. Porter, J.E. Halpert, T.S. Mentzel, M.A. Kastner, M.G. Bawendi, Charge transport in mixed CdSe and CdTe colloidal nanocrystal films. Phys. Rev. B: Condens. Matter Mater. Phys. **82** (2010). https://doi.org/10.1103/PhysRevB.82.155201

51. O. Madelung, U. Rössler, M. Schulz (eds.), in *II-VI I-VII Compd. Semimagn. Compd.* Cadmium telluride (CdTe) effective masses (Springer, Berlin, 1999), pp. 1–2. https://doi.org/10.1007/10681719_625

52. G.G. Roberts, M.C. Petty, I.M. Dharmadasa, Photovoltaic properties of cadmium-telluride/Langmuir-film solar cells. IEEE Proc. I Solid State Electron Devices. **128**, 197 (1981). https://doi.org/10.1049/ip-i-1.1981.0049

53. T.Y. Chang, C.L. Chang, H.Y. Lee, P.T. Lee, A metal-insulator-semiconductor solar cell with high open-circuit voltage using a stacking structure. IEEE Electron Device Lett. **31**, 1419–1421 (2010). https://doi.org/10.1109/LED.2010.2073437

54. S. Chander, A. Purohit, A. Sharma, S.P. Nehra, M.S. Dhaka, Impact of temperature on performance of series and parallel connected mono-crystalline silicon solar cells. Energy Reports. **1**, 175–180 (2015). https://doi.org/10.1016/j.egyr.2015.09.001

55. W.L. Liu, Y.L. Chen, A.A. Balandin, K.L. Wang, Capacitance–voltage spectroscopy of trapping states in GaN/AlGaN heterostructure field-effect transistors. J. Nanoelectron. Optoelectron. **1**, 258–263 (2006). https://doi.org/10.1166/jno.2006.212.

56. S.W. Lin, J. Du, C. Balocco, Q.P. Wang, A.M. Song, Effects of bias cooling on charge states in heterostructures embedding self-assembled quantum dots. Phys. Rev. B **78**, 115314 (2008). https://doi.org/10.1103/PhysRevB.78.115314

57. I. Strzalkowski, S. Joshi, C.R. Crowell, Dielectric constant and its temperature dependence for GaAs, CdTe, and ZnSe. Appl. Phys. Lett. **28**, 350–352 (1976). https://doi.org/10.1063/1.88755

58. D. Neamen, Semiconductor physics and devices. Mater. Today **9**, 57 (2006). https://doi.org/10.1016/S1369-7021(06)71498-5

59. N.B. Chaure, S. Bordas, A.P. Samantilleke, S.N. Chaure, J. Haigh, I.M. Dharmadasa, Investigation of electronic quality of chemical bath deposited cadmium sulphide layers used in thin film photovoltaic solar cells. Thin Solid Films **437**, 10–17 (2003). https://doi.org/10.1016/S0040-6090(03)00671-0

Chapter 4
ZnS Deposition and Characterisation

4.1 Introduction

Zinc sulphide (ZnS) from the II–VI groups has been well explored for its physical, chemical and electronic properties. The optoelectronic and photovoltaic properties of this material are continually drawing more attention due to its extensive application in light-emitting devices and sensors and as buffer or window layers in photovoltaic solar cell configurations [1, 2]. Although the use of cadmium sulphide (CdS) as a window layer in a cadmium telluride (CdTe)-based solar cell configuration has yielded high photovoltaic conversion efficiency [3, 4], the use of CdS is constrained due to the parasitic absorption in CdS below 520 nm photon wavelengths [5] as a result of its 2.42 eV bandgap. ZnS, on the other hand, possesses a wider bandgap of ~3.70 eV and has consequently been explored as a substitute for CdS [6], but factors such as the high lattice mismatch between ZnS and CdTe, the comparatively high conduction band edge of ZnS as compared to CdTe and the high resistivity of ZnS have necessitated the retention of CdS in the ZnS/CdS/CdTe solar cell configuration. However, importantly, the inclusion of ZnS provides good wetting property and allows for the reduction of CdS thickness to be reduced without an increase in the shunting paths [7] and hence reduces the parasitic absorption of CdS [5].

Over ten different techniques have been recorded in the literature for possible growth of ZnS such as metalorganic chemical vapour deposition [8], spray pyrolysis [9] and electrodeposition [10] amongst others. The electrodeposition technique has potential advantages when compared to others with attributes such as low cost, simplicity, scalability and the manufacturability across large surface area [11].

Literature records that the electrodeposition of crystalline ZnS has been achieved mainly by the inclusion of complexing agents such as ammonia (NH_3) [10], hydrazine (N_2H_4) [12], trisodium citrate ($Na_3C_6H_5O_7$) [13] and glycerol ($C_3H_8O_3$) [14] in the electrolytic bath. The literature on the electrodeposition of ZnS without a complexing agent has been scarce, and the reported ZnS layers were structurally amorphous [15, 16]. Hence, with a view towards cost reduction and process

© Springer International Publishing AG, part of Springer Nature 2019
A. A. Ojo et al., *Next Generation Multilayer Graded Bandgap Solar Cells*,
https://doi.org/10.1007/978-3-319-96667-0_4

simplification, this chapter presents in full the electrodeposition process and characterisation of crystalline ZnS without the use of binding/complexing agent in a two-electrode configuration.

4.2 Electrolytic Bath and Substrate Preparation for ZnS

4.2.1 Electrolytic Bath Preparation

ZnS thin films were electrodeposited cathodically on glass/FTO substrates by the potentiostatic technique in which the counter electrode (anode) was a high-purity graphite rod. Zinc sulphate monohydrate ($ZnSO_4 \cdot H_2O$) of 99.9% purity and ammonium thiosulphate (($NH_4)_2S_2O_3$) of 98% purity were used as the zinc and sulphur precursors, respectively. The electrolytic bath was prepared by dissolving 0.2 M $ZnSO_4 \cdot H_2O$ and 0.2 M ($NH_4)_2S_2O_3$ in 400 mL deionised (DI) water contained in a 500 mL polypropylene beaker. The polypropylene beaker was placed in an 800 mL glass beaker containing DI water. The DI water contained in the outer glass beaker helps to maintain uniform heating of the electrolyte. The solution was continuously stirred and electro-purified at 1000 mV for ~100 h to achieve complete dissolution and also to reduce the level of impurity. Following deposition, ZnS layers electrodeposited at different cathodic voltages were characterised. For this set of experiments, the bath temperature was maintained at 30 °C, and the pH value of the bath was set to 4.00 ± 0.02 using dilute H_2SO_4 and NH_4OH at the start of deposition. It is important to maintain the pH of the electrolytic bath between 3.00 and 4.50 as an increase or decrease in pH outside this range results in the formation of white precipitates of sulphur and rapid deposition of ZnS layers with low adhesion to the underlying glass/FTO layer.

In this study, the two-electrode electrodeposition configuration was utilised for simplicity. A glass/FTO substrate and a high-purity carbon electrode serve as the working and counter electrodes, respectively. The working electrode of sheet resistance ~7 Ω/sq (TEC7) was cut into 3×2 cm^2 sections and cleaned ultrasonically before deposition. The electrodeposition process was performed using a computerised Gill AC potentiostat. Prior to the deposition of ZnS layers, cyclic voltammograms of the resulting electrolyte(s) was performed to determine the possible deposition voltage range of ZnS.

4.2.2 Substrate Preparation

The substrates were ultrasonically cleaned for 20 min in an ultrasonic bath containing soap solution and subsequently rinsed in deionised (DI) water. The substrates were degreased using both methanol and acetone and rinsed in DI water. Finally, the glass/FTO layers were submerged in a clean beaker of DI water and transferred directly into the electroplating bath. Prior to characterisation, the electroplated ZnS layers were rinsed, dried and divided into two halves. Half were left as-deposited, and the other half were heat treated at 300 °C for 10 min in air to enhance both their material and electronic properties [16]. Both the as-deposited and the heat-treated ZnS layers were subsequently characterised for their structural and electronic properties.

4.3   Growth and Voltage Optimization of ZnS

ZnS layers were electrodeposited between the deposition voltages 1350 and 1500 mV for 2 h per sample. The deposition range was determined based on information obtained via cyclic voltammetry as later discussed in Sect. 4.3.1.

4.3.1   Cyclic Voltammetric Study

Figure 4.1 shows a typical cyclic voltammogram of an aqueous solution of 0.2 M $ZnSO_4 \cdot H_2O$ and 0.2 M $(NH_4)_2S_2O_3$ in 400 mL of DI water during the forward and reverse cycle between 0 and 2000 mV cathodic voltage. The scanning rate was fixed at 3 mVs^{-1} with the bath temperature, stirring rate and pH set to 30 °C, 300 rpm and 4.00 ± 0.02, respectively.

The growth temperature was set to 30 °C to avoid vigorous sulphur precipitation which occurs at a higher temperature and amorphous phase ZnS which occurs at a lower temperature. Based on the standard reduction potential values of both Zn^{2+} and S^{2-} with $E° = -0.762$ V and $E° = 0.449$ V, respectively, with respect to standard H_2 electrode, sulphur deposits first at ~300 mV, followed by the deposition

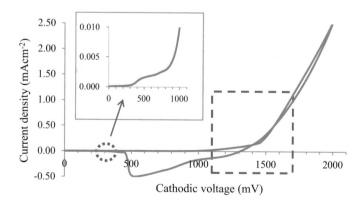

Fig. 4.1 A typical cyclic voltammogram for deposition electrolyte containing the mixture of 0.2 M $ZnSO_4 \cdot H_2O$ and 0.2 M $(NH_4)_2S_2O_3$ in 400 mL of DI water at ~30 °C and pH 4.00 ± 0.02. The inset is the expanded forward cycle between 0 and 1100 mV

of zinc at ~1100 mV in the forward cycle. On the reverse cycle, the dissolution of zinc starts at ~1300 mV, followed by sulphur at ~540 mV. It was deduced that between the cathodic voltage range of 1100 and 1700 mV, ZnS can be achieved as either S-rich, stoichiometric or Zn-rich ZnS. A stable current density of ~150 mAcm^{-2} is observable between 1320 and 1500 mV cathodic voltages. ZnS layers were grown within this range at the step of 50 mV and characterised to identify the optimum growth voltage in which stoichiometric/near-stoichiometric ZnS can be achieved. On this initial experimentation using only XRD analysis, crystalline ZnS was observed only at 1420 mV. Therefore, surrounding cathodic potentials were scanned at the steps of 10 mV to identify the optimum growth voltage. The final electrochemical equation for the formation of ZnS at the cathode can be written as Equation 4.1:

$$Zn^{2+} + 2e^- + S \rightarrow ZnS \qquad \text{(Equation 4.1)}$$

4.3.2 X-Ray Diffraction Study

A typical XRD analysis for both as-deposited and heat-treated ZnS at 300 °C for 10 min is shown in Fig. 4.2a, b, respectively. As observed in Fig. 4.2a, cathodic voltages outside the range of 1410-1430 mV show an amorphous behaviour which might be as a result of the growth technique, the cathodic voltage, the growth temperature and the absence of binding/complexing agent in the electrolytic bath. For the ZnS layers grown at 1430 mV, peaks associable with ZnS (111)C, ZnS (110)H and ZnS (220)C were observed at $2\theta = $ ~29.2°, ~47.5° and ~48.2° respectively. Conversely, the ZnS layers grown at 1420 mV show peaks associable with ZnS (101)H, ZnS (002)H at $2\theta = $ ~30.8° and (002)H at 32.16° respectively. An inclusion of ZnO (101)H and unreacted Zn (101)H phase were also observed at $2\theta = 36.18°$ and 42.88° for most of the deposited layers across the explored cathodic voltage range. The presence of ZnO might be as a result of the aqueous solution utilised and/or the oxidisation of the ZnS surface in air.

With regard to the XRD spectra of heat-treated ZnS layers shown in Fig. 4.2b, amorphisation of all the ZnS layers grown at all the cathodic voltages explored was observed except for layers grown at 1430 mV. A retention of the ZnS (111)C was observed at $2\theta = 29.22°$ but with reduced peak intensity after heat treatment (see Fig. 4.2b).

The amorphisation of the crystalline ZnS grown at 1430 mV might be due to a high defect concentration and the deterioration of the ZnS layer [14], elemental composition, oxidation, and the sublimation of the layer leading to the degradation of the ZnS layer. The crystallite size was calculated using the Scherrer's formula (see Equation 3.8), and the XRD data is summarised in Table 4.1. Based on the observation in Table 4.1, the highest crystallite size of 66.1 nm was observed for the as-deposited ZnS layers grown at 1430 mV. After heat treatment, a reduction in the crystallite size along the same plane was observed for the ZnS layers grown at 1430 mV due to randomisation and the oxidation of the ZnS layer [14]. This

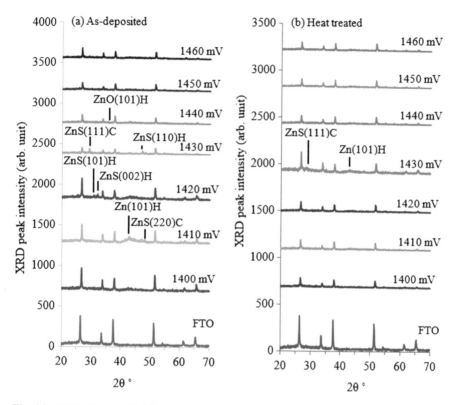

Fig. 4.2 XRD patterns of ZnS layers grown between 1400 and 1460 mV cathodic voltage for (**a**) as-deposited and (**b**) heat-treated ZnS layers at 300 °C for 10 min

Table 4.1 The XRD analysis of ZnS layers grown at cathodic voltage of 1420 and 1430 mV for the as-deposited and the heat-treated layers at 300 °C for 10 min in air

Sample	2θ (°)	d-spacing (Å)	FWHM (°)	Crystallite size (nm)	Plane (hkl)	Assignments
AD						
1420 mV	32.16	2.80	0.2598	33.3	(200)	ZnS cubic
1430 mV	29.22	3.05	0.1299	66.1	(111)	ZnS cubic
HT						
1430 mV	28.62	3.12	0.3031	28.3	(111)	ZnS cubic

observation can be attributed to the re-crystallisation of ZnS resulting in the competing phases of hexagonal and cubic ZnS. Another possibility is the oxidation of ZnS surface to give ZnO with competing crystalline phases resulting to collapse of the ZnS crystallinity. The extracted XRD data from this ZnS material work matches the Joint Committee on Powder Diffraction Standards (JCPDS) reference file number 00-05-566, 00-05-492 and 00-36-1450 for ZnS, 00-036-1451 for ZnO and 00-004-0831 for Zn. The presence of the narrow crystalline range of electrodeposited ED-ZnS and the amorphisation of the ED-ZnS layers is very different to the ZnS layers electroplated from the bath incorporating complexing agents [10, 12, 13] with improvement in the ZnS layer crystallinity with the increasing complexing agent concentration.

4.3.3   Raman Study

Figure 4.3a, b shows the Raman spectra of ~600-nm-thick ZnS layers grown at 1420 mV, 1430 mV and 1440 mV cathodic voltages on glass/FTO substrates under as-deposited and heat-treated condition, respectively. Raman peaks associated with ZnS were observed at 147 and 217 cm^{-1} which are identified as 2TA and 2LA optical mode phonons for ZnS [17, 18] under both as-deposited and heat-treated conditions.

With respect to the as-deposited ZnS layers (Fig. 4.3a), sharp Raman peaks were observed at 147 and 217 cm^{-1} for the layers grown at 1420 and 1430 mV, while broad peaks were observable for the ZnS layers grown at 1440 mV. It should be noted that the ZnS layers grown at 1440 mV is similar to all other explored ZnS layers grown at 1400–1460 mV with the exception of 1420 and 1430 mV grown layers. As documented in the literature [19, 20], the broadness of the observed peaks suggests the amorphous nature of the most of the ZnS layers which are in accord with the summations made in the XRD section (see Sect. 4.3.2). The broadening of the peaks observed at 147 and 217 cm^{-1} of the ZnS layers grown at 1420 and 1430 mV shows the amorphisation of the layers due to the deterioration of the layers, most probably due to oxidation.

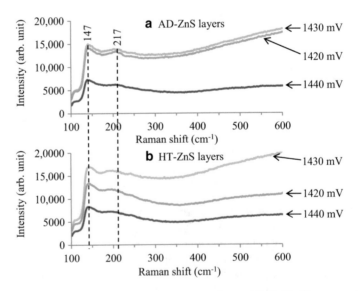

Fig. 4.3 Raman spectra of (**a**) as-deposited and (**b**) heat-treated ZnS thin films grown between 1420 and 1440 mV

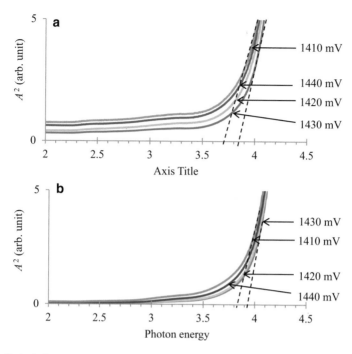

Fig. 4.4 Optical absorption spectra for electrodeposited ZnS thin films between voltage range of 1410 and 1440 mV for (**a**) as-deposited and (**b**) heat-treated ZnS at 300 °C for 10 min in air

4.3.4   Optical Property Analyses

Figure 4.4a shows the optical absorbance results for ~200-nm-thick ZnS layers grown between the 1410 and 1440 mV cathodic voltages under the as-deposited condition, and Fig. 4.4b shows the equivalent after heat treatment at 300 °C for 10 min. The absorbance squared (A^2) was plotted against the energy of photons, and the extrapolated straight-line section of the graph to the energy of photon axis (at $A^2 = 0$) gives an estimate of the bandgap energy. With respect to the as-deposited ZnS layers grown between 1410 and 1440 mV (Fig. 4.4a), the observed bandgap ranges between 3.70 and 3.85 eV. After heat treatment, an increase in the observed bandgap ≥ 3.84 eV and a reduction in the bandgap range between 3.84 and 3.94 eV were observed. The deviation of the observed bandgap energy from that of the bulk cubic-ZnS of ~3. 70 eV [1] might be due to the incorporation of hexagonal ZnS which is known to have a bandgap of ~3.90 eV (see Section 4.3.2).

Figure 4.5a shows the optical transmittance of 150-nm-thick ZnS layers grown between 1410 and 1440 mV under as-deposited and in Fig. 4.5b after heat treatment at 300 °C for 10 min. Comparatively, a slight increase in transmittance was observed for all the explored ZnS layers after heat treatment. It is noteworthy that all the samples show transmittance above 50% for wavelengths larger than 335 nm, with the highest transmittance of ~80% observed for the layers grown at 1420 and

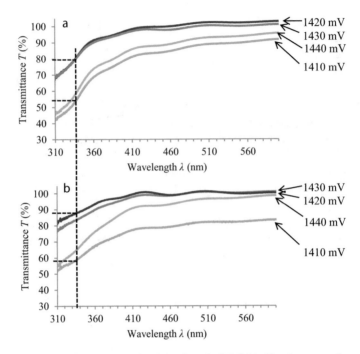

Fig. 4.5 Optical transmittance spectra for electrodeposited ZnS thin films between voltage range of 1410 and 1440 mV for (**a**) as-deposited and (**b**) heat-treated ZnS at 300 °C for 10 min in air

1430 mV. The ZnS layers grown at 1420 and 1430 mV show good optical properties suitable for a buffer layer in a photovoltaic device configuration with negligible parasitic absorbance of incident photons to the cell.

4.3.5   Morphological Studies

Figure 4.6a–c shows SEM micrographs of as-deposited ZnS layers grown at 1410 mV, 1430 mV and 1440 mV, respectively, while Fig. 4.6d–f shows their respective heat-treated micrographs. For this set of experiments, the ZnS layers utilised were ~200 nm thick grown on glass/FTO. For the explored cathodic voltage range between 1410 and 1440 mV under both the as-deposited and heat-treated conditions, full coverage of the underlying glass/FTO substrate was observed. A slight increase in the grain sizes of the as-deposited ZnS layers ranging from ~(100–380) nm to ~(120–400) nm after heat treatment was observable. Further to this, it appears that the ZnS layers grown at 1430 and 1440 mV coalesce after heat treatment.

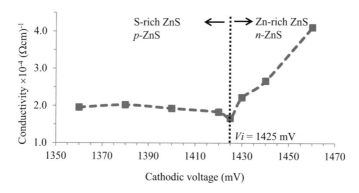

Fig. 4.8 A typical graph of electrical conductivity against growth voltage of heat-treated ZnS layers

a gradual reduction in the electrical conductivity value from the ZnS layers grown at 1360 mV towards the ZnS layer grown at 1425 mV is observed in Fig. 4.8. Conversely, a gradual increase in the electrical conductivity was observed with increasing cathodic voltage from which the ZnS layers were grown.

In relation to the cathodic voltage-dependent deposition sequence as discussed in Sect. 4.3.1, the observed trend in Fig. 4.8 can be attributed to the deposition of S-rich ZnS layers at cathodic voltages lower than ~1425 mV and the increasing Zn richness of the ZnS layers grown above 1425 mV. It is clear that the ZnS layers grown at the 1425 mV range show relatively lowest conductivity due to the stoichiometric and the intrinsic nature of the ZnS layers grown since practically all valence electron are engaged in bonding.

4.4 Conclusions

The work presented in this chapter demonstrates the successful deposition of both p- and n-type ZnS thin films from an aqueous solution containing 0.2 M $ZnSO_4 \cdot H_2O$ and 0.2 M $(NH_4)_2S_2O_3$ using the two-electrode configuration. Based on the selected material characterisation techniques explored, 1430 mV was identified as the best cathodic potential in which stoichiometric ED-ZnS is achievable. The XRD results show that crystalline ZnS can be electrodeposited at 1420 and 1430 mV but are amorphized after 300 °C heat treatment. SEM imaging shows small grain structure attesting to the wetting property of ZnS. The optical bandgap of the ED-ZnS layers is within the reported values for bulk hexagonal ZnS layers (~3.90 eV).

References

1. T. Nakada, M. Mizutani, 18% Efficiency Cd-free Cu(In, Ga)Se 2 thin-film solar cells fabricated using chemical bath deposition (CBD)-ZnS buffer layers. Jpn. J. Appl. Phys. **41**, L165 (2002). http://stacks.iop.org/1347-4065/41/i=2B/a=L165

2. S.M. Sze, K.K. Ng, *Physics of Semiconductor Devices* (Wiley, Hoboken, 2006). https://doi.org/10.1002/0470068329

3. J. Britt, C. Ferekides, Thin-film CdS/CdTe solar cell with 15.8% efficiency. Appl. Phys. Lett. **62**, 2851–2852 (1993). https://doi.org/10.1063/1.109629

4. T.L. Chu, S.S. Chu, C. Ferekides, C.Q. Wu, J. Britt, C. Wang, 13.4% Efficient thin-film CdS/CdTe solar cells. J. Appl. Phys. **70**, 7608 (1991). https://doi.org/10.1063/1.349717.

5. J.E. Granata, J.R. Sites, in *Conf. Rec. Twenty Fifth IEEE Photovolt. Spec. Conf. 1996*. Effect of CdS thickness on CdS/CdTe quantum efficiency (2000), pp. 853–856. https://doi.org/10.1109/PVSC.1996.564262

6. A. Ennaoui, W. Eisele, M. Lux-Steiner, T.P. Niesen, F. Karg, Highly efficient Cu(Ga,In)(S,Se)2 thin film solar cells with zinc-compound buffer layers. Thin Solid Films. **431–432**, 335–339 (2003). https://doi.org/10.1016/S0040-6090(03)00155-X

7. A.A. Ojo, H.I. Salim, O.I. Olusola, M.L. Madugu, I.M. Dharmadasa, Effect of thickness: a case study of electrodeposited CdS in CdS/CdTe based photovoltaic devices. J. Mater. Sci. Mater. Electron. **28**, 3254–3263 (2017). https://doi.org/10.1007/s10854-016-5916-0

8. J. Hu, G. Wang, C. Guo, D. Li, L. Zhang, J. Zhao, Au-catalyst growth and photoluminescence of zinc-blende and wurtzite ZnS nanobelts via chemical vapor deposition. J. Lumin. **122–123**, 172–175 (2007). https://doi.org/10.1016/j.jlumin.2006.01.074

9. A.N. Yazici, M. Öztaş, M. Bedir, Effect of sample producing conditions on the thermoluminescence properties of ZnS thin films developed by spray pyrolysis method. J. Lumin. **104**, 115–122 (2003). https://doi.org/10.1016/S0022-2313(02)00686-5

10. E.M. Mkawi, K. Ibrahim, M.K.M. Ali, M.A. Farrukh, A.S. Mohamed, Electrodeposited ZnS precursor layer with improved electrooptical properties for efficient Cu2ZnSnS4 thin-film solar cells. J. Electron. Mater. **44**, 3380–3387 (2015). https://doi.org/10.1007/s11664-015-3849-7

11. I.M. Dharmadasa, *Advances in Thin-Film Solar Cells* (Pan Stanford, Singapore, 2013)

12. S. Tec-Yam, J. Rojas, V. Rejón, A.I. Oliva, High quality antireflective ZnS thin films prepared by chemical bath deposition. Mater. Chem. Phys. **136**, 386–393 (2012). https://doi.org/10.1016/j.matchemphys.2012.06.063.

13. G.L. Agawane, S.W. Shin, A.V. Moholkar, K.V. Gurav, J.H. Yun, J.Y. Lee, J.H. Kim, Non-toxic complexing agent Tri-sodium citrate's effect on chemical bath deposited ZnS thin films and its growth mechanism. J. Alloys Compd. **535**, 53–61 (2012). https://doi.org/10.1016/j.jallcom.2012.04.073

14. B.W. Sanders, A.H. Kitai, The electrodeposition of thin film zinc sulphide from thiosulphate solution. J. Cryst. Growth. **100**, 405–410 (1990). https://doi.org/10.1016/0022-0248(90)90238-G

15. O.K. Echendu, I.M. Dharmadasa, Effects of thickness and annealing on optoelectronic properties of electrodeposited ZnS thin films for photonic device applications. J. Electron. Mater. **43**, 791–801 (2013). https://doi.org/10.1007/s11664-013-2943-y

16. M.L. Madugu, O.I.-O. Olusola, O.K. Echendu, B. Kadem, I.M. Dharmadasa, Intrinsic doping in electrodeposited ZnS thin films for application in large-area optoelectronic devices. J. Electron. Mater. **45**, 2710–2717 (2016). https://doi.org/10.1007/s11664-015-4310-7

17. A. Fairbrother, V. Izquierdo-Roca, X. Fontané, M. Ibáñez, A. Cabot, E. Saucedo, A. Pérez-Rodríguez, ZnS grain size effects on near-resonant Raman scattering: optical non-destructive grain size estimation. CrystEngComm. **16**, 4120 (2014). https://doi.org/10.1039/c3ce42578a

18. Y. Ebisuzaki, M. Nicol, Raman spectrum of hexagonal zinc sulfide at high pressures. J. Phys. Chem. Solids. **33**, 763–766 (1972). https://doi.org/10.1016/0022-3697(72)90088-1

19. H. Richter, Z.P. Wang, L. Ley, The one phonon Raman spectrum in microcrystalline silicon. Solid State Commun. **39**, 625–629 (1981). https://doi.org/10.1016/0038-1098(81)90337-9

20. I.H. Campbell, P.M. Fauchet, The effects of microcrystal size and shape on the one phonon Raman spectra of crystalline semiconductors. Solid State Commun. **58**, 739–741 (1986). https://doi.org/10.1016/0038-1098(86)90513-2

21. A.B. Bhalerao, C.D. Lokhande, B.G. Wagh, Photoelectrochemical cell based on electrodeposited nanofibrous ZnS thin film. IEEE Trans. Nanotechnol. **12**, 996–1001 (2013). https://doi.org/10.1109/TNANO.2013.2272469

Chapter 5
CdS Deposition and Characterisation

5.1 Introduction

The study of the structural, optical, morphological and physical properties of binary compound semiconductors such as cadmium sulphide (CdS) thin films is a subject of current interest due to its applications in optoelectronic and large-area electronic devices. In photovoltaics, polycrystalline CdS thin films are often used as a window layer in the CdS/CdTe solar cell configuration to achieve high conversion efficiency. CdS thin films have been grown using over ten different techniques [1] as reported in the literature with electrodeposition edging other deposition based on its simplicity, low cost and scalability amongst other attributes [2]. The electrodeposition of CdS as reported in the literature is done mainly by using sodium thiosulfate ($Na_2S_2O_3$) as the sulphur precursor. This precursor is associated with the formation of sulphur precipitates during growth and the accumulation of sodium, Na, in the electrolytic bath [3]. The incorporation of p-type dopants such as Na into CdS layers through absorption or chemical reaction will reduce the electrical conductivity of the grown layer and thus constitute a drawback in the electrical property of the CdS layer [3]. Therefore, in view of depositing CdS from other sulphur precursors without the aforementioned drawbacks, other sulphur sources need to be explored. Sulphur source such as thiourea (NH_2CSNH_2) have been well established in the growth of CdS using chemical bath deposition (CBD) technique, but the literature on the use of the electrochemical technique for this precursor is scarce. The electrodeposition (ED) technique is comparatively advantageous with respect to deposition process continuity and Cd-containing waste reduction. Although, preliminary investigation of the electrodeposition of CdS from thiourea was explored in 2001 by Yamaguchi et al. [4], this chapter presents in full the comprehensive details of the growth and characterisation of CdS from NH_2CSNH_2 precursor and further investigates the effect of $CdCl_2$ treatment, film thickness and heat treatment duration on the electronic quality of electrodeposited CdS layers.

© Springer International Publishing AG, part of Springer Nature 2019
A. A. Ojo et al., *Next Generation Multilayer Graded Bandgap Solar Cells*,
https://doi.org/10.1007/978-3-319-96667-0_5

This chapter goes beyond the growth and characterisation norm into the comparative analysis of the electronic properties of electrodeposited CdS using thiourea precursor to CBD-CdS and single-crystal CdS previously reported in the literature.

5.2   Electrolytic Bath and Substrate Preparation for CdS

5.2.1   Electrolytic Bath Preparation

CdS thin-films were cathodically electrodeposited on glass/FTO substrates by the potentiostatic technique in which the counter electrode was a high purity graphite rod. Cadmium chloride hydrate ($CdCl_2 \cdot xH_2O$) of 98% purity and thiourea (NH_2CSNH_2) of 99% purity were used as cadmium and sulphur sources, respectively. The electrolyte was prepared by dissolving 0.12 M $CdCl_2 \cdot xH_2O$ and 0.18 M NH_2CSNH_2 in 800 mL deionised (DI) water contained in a 1000 mL polypropylene beaker. The polypropylene beaker was placed in an 1800 mL glass beaker containing DI water. The glass beaker serves as the outer bath and helps to maintain uniform heating of the electrolyte. The ED bath described is a mimic of the setup as discussed in Sect. 3.3. The solution was stirred and electro-purified for ~50 h to reduce the impurity level and to achieve homogeneity of the solution. The electrolytic bath containing 0.12 M $CdCl_2 \cdot xH_2O$ and 0.18 M NH_2CSNH_2 in 800 mL DI water will be referred to as CdS bath henceforth.

The cyclic voltammograms of the resulting electrolyte(s) were recorded before the deposition of CdS layer to determine the possible deposition voltage range of CdS. A complete characterisation of the CdS grown at different voltages in the selected range was undertaken to determine the optimum growth voltage (V_g). For these experiments, the bath temperature was maintained at 85 °C during the CdS growth to achieve higher crystallinity due to high deposition temperature [5]. However, the temperature increase is limited due to the use of aqueous solution. The pH value was adjusted to 2.50 ± 0.02 at the start of deposition using diluted solutions of HCl and NH_4OH for all the samples. It is essential to maintain the pH of the electrolytic bath between 2.00 and 3.00 as an increase or decrease in pH outside this range results into the formation of white precipitates of cadmium hydroxide and rapid precipitation of CdS, respectively [6]. The two-electrode configuration was used in this study with glass/FTO as the working electrode. The working electrodes of sheet resistance ~7 Ω/sq were cut into 2 × 4 cm^2 pieces and a high purity carbon rod was used as the anode. A computerised Gill AC potentiostat was utilised as the power supply source.

5.2.2   Substrate Preparation

Substrates were ultrasonically cleaned at the initial stage in soap solution for 20 min and rinsed in deionised (DI) water. The substrates were then cleaned thoroughly with methanol and acetone to remove any grease and rinsed in DI water. Finally, the FTO is submerged in a clean beaker of DI water and transferred directly into the electroplating bath. Prior to characterisation, electrodeposited CdS layers were rinsed, dried and divided into two halves. Half were left as-deposited, and the other half were CdCl$_2$ treated at 400 °C in the air to enhance both their material and electronic properties. The CdCl$_2$ treatment was performed by adding few drops of aqueous solution containing 0.1 M CdCl$_2$ in 20 mL of DI water to the surface of the semiconductor layer. The full coverage of the layers with the treatment solutions was achieved by spreading the solution using solution-damped cotton bud. The semiconductor layer was allowed to air-dry and heat treated at 400 °C for 20 min. Both the as-deposited and CdCl$_2$-treated CdS layers were characterised afterwards for both their material and electronic properties.

5.3   Growth and Voltage Optimization of CdS

Aside from the glass/FTO substrate utilised for cyclic voltammetry, the CdS layers were electrodeposited between the deposition voltages 789 and 793 mV for 2 h per sample. The deposition range was determined based on information obtained via cyclic voltammetry (see Sect. 5.3.1).

5.3.1   Cyclic Voltammetric Study

The cyclic voltammogram is a plot of current density as a function of the applied voltage across an electrolytic bath. Figure 5.1 shows the cyclic voltammogram of an aqueous solution of a mixture of 0.12 M CdCl$_2$·xH$_2$O and 0.18 M NH$_2$CSNH$_2$ in 800 mL of DI water during the forward and reverse cycles between 100 and −2000 mV cathodic voltage at pH 2.50 ± 0.02 and a fixed scan rate of 3 mV s^{-1}. The stirring rate and temperature of the bath were kept constant at 300 rpm and 85 °C, respectively.

  According to the redox potential value of cadmium and sulphur ions, sulphur deposits first (with an $E°$ value of −0.43 V w.r.t. standard H$_2$ electrode), followed by cadmium (with an $E°$ value of −0.40 V w.r.t. standard H$_2$ electrode). The complete electrochemical equation for the formation of CdS at the cathode can be written as Equation 5.1.

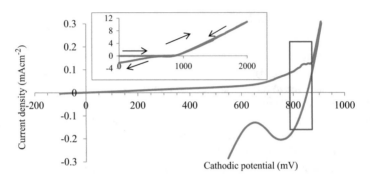

Fig. 5.1 A typical cyclic voltammogram for deposition electrolyte containing the mixture of 0.12 M CdCl$_2$·xH$_2$O and 0.18 M NH$_2$CSNH$_2$ at ~85 °C and pH = 2.50 ± 0.02. The inset shows the full cyclic voltammogram measured between 0 and 2000 mV

$$Cd^{2+} + S_2O_3^{2-} + 2e^- \rightarrow CdS + SO_3^{2-} \qquad \text{(Equation 5.1)}$$

It was observed in the forward cycle that between the cathodic voltage range 785 and 880 mV, the deposition current density appears to be fairly stable within the range of ~130 μA cm^{-2}. The growth voltage range (i.e. 785–880 mV) had been pre-characterised at cathodic voltage steps of 10 mV using XRD analysis (these preliminary results are not presented in this book). The highest XRD peak intensity was observed at 790 mV. Therefore, the surrounding cathodic voltage was scanned at steps of 1 mV to investigate surrounding growth voltage values further. The result of the characterised CdS layers grown between the cathodic voltages of (788–793) mV using PEC cell measurement, optical absorbance and transmittance measurement, XRD and SEM were used to determine the optimum growth conditions.

It was also observed during the voltammetric cycle that the deposited film colour changes to transparent yellow at growth voltages (V_g) ≤780 mV which suggests a region of sulphur richness. The deposited film colour turns to greenish-yellow at growth voltage above 780 mV which indicate the presence of stoichiometric or near-stoichiometric CdS thin films. As the growth voltage increases above 793 mV, the film colour changes to dark green as shown in Fig. 5.2. This can be attributed to Cd richness in CdS. Above 880 mV cathodic voltage, high-current density is observed with a detrimental effect on the ED-CdS layer quality. The sharp increase in current density might be due to the electrolysis of water and the deposition of Cd dendrites.

Based on observation, the stoichiometric CdS thin-film range can be qualitatively predicted by appearance during the voltammetric cycle. But, exploring other material characterisation techniques is still required for quantitative assessment of the deposited CdS layers at different cathodic voltages. It was interesting to observe no sulphur precipitation throughout these sets of experiments which is one of the problems associated with other sulphur precursors in the electrodeposition of CdS [7, 8].

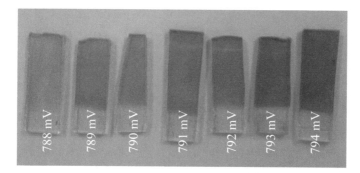

Fig. 5.2 The physical appearances of as-deposited CdS grown at V_g from 788 to 794 mV grown for 20 min each

5.3.2 X-Ray Diffraction Study

After preliminary growth and characterisation, CdS layers were grown between 788 and 793 mV cathodic voltages at 1 mV step change so as to comb the cathodic voltages surrounding the predetermined 790 mV to identify the stoichiometric or near-stoichiometrically grown CdS layer. For this work, each CdS samples were grown for the same time duration and $CdCl_2$ treated under the same condition. The observed XRD peaks/crystallinity for both the as-deposited and $CdCl_2$-treated CdS layers within the cathodic voltage range of 788–793 mV were compared as shown in Figs. 5.3a, b, 5.4 and 5.5.

As observed in Fig. 5.3, all the as-deposited CdS layers were polycrystalline with a preferred peak orientation along the CdS (002)H plane which coincides with the FTO peak at angle $2\theta = 26.68°$. The as-deposited CdS layers were polycrystalline with peak orientations corresponding to CdS (100)H at $2\theta = 24.90°$, CdS (002)H at $2\theta = 26.68°$, CdS (101)H at $2\theta = 28.86°$, CdS (220)C at $2\theta = 46.98°$, CdS (103)H at 48.01°, Cd (101)H at $2\theta = 38.89°$ and S (319)O at $2\theta = 42.63°$ present. This indicates the inclusion of elemental Cd and S, together with cubic CdS phase within the main phase of hexagonal CdS. The preferred peak orientation along the CdS (002) H plane coincides with FTO peak at $2\theta = 26.68°$. Due to the effect of the FTO peak on the CdS (002)H peak intensity, CdS (101)H at $2\theta = 28.86°$ with the second predominant peak intensity was utilised as the preferred peak for analysis of the CdS layers. After $CdCl_2$ treatment, peaks identified as Cd (101)H at $2\theta = 38.89°$, S (319)O at $2\theta = 42.63°$ and CdS (220)C at $2\theta = 46.98°$ completely disappears as observed in Figs. 5.3b and 5.5. This can be attributed to the diffusion and evaporation of excess sulphur, the formation of CdS from the reaction between elemental Cd and S to form hexagonal-CdS and the instability of cubic phase CdS after heat treatment [9, 10]. A substantial increase in XRD peak intensity of CdS in hexagonal phases was also observed after $CdCl_2$ treatment except for CdS (103)H as shown in Fig. 5.5. After $CdCl_2$ treatment, the dark colour (Cd-richness) disappears which is indicative of the evaporation of excess elemental Cd or its conversion into CdS or CdS. Since the CdO has a bandgap of 2.35 eV [11], it shows similar optical properties to that of CdS.

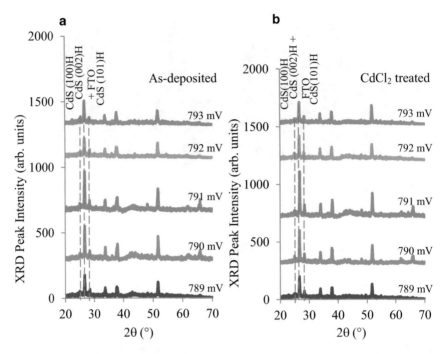

Fig. 5.3 Typical XRD patterns of CdS layers grown between 789 and 793 mV deposition potential for (**a**) as-deposited CdS layers and (**b**) CdCl₂-treated CdS layers at 400 °C for 20 min

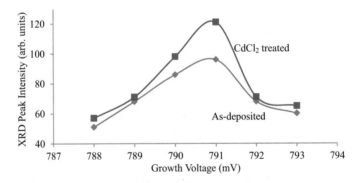

Fig. 5.4 Comparative analysis of CdS (101)H peak for as-deposited and CdCl₂-treated layer grown at different growth voltages between 788 and 793 mV

It is well documented in the literature that CdS can grow in two different crystalline structures, namely, the hexagonal (wurtzite structure) and the cubic (zinc blend structure), with the hexagonal being the more metastable phase [9, 10]. Hence, the hexagonal CdS phases were retained after CdCl₂ treatment at 400 °C for 20 min. The initial presence of CdS in both hexagonal and cubic phases in the as-deposited layers may be attributed to vigorous stirring during growth at low

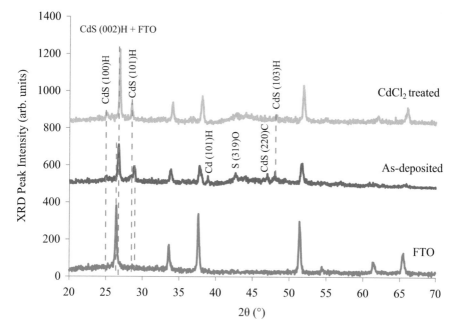

Fig. 5.5 Comparative analysis of XRD peaks from as-deposited and CdCl$_2$-treated CdS layers grown at 791 mV

temperature as argued by Kaur et al. [8]. All the hexagonal phases of CdS in the as-deposited layer showed a substantial increase in intensity after CdCl$_2$ treatment except for CdS (103)H as shown in Fig. 5.5. These alterations in the XRD peak intensity might be due to the reorientation of the crystal lattice during heat treatment in the presence of CdCl$_2$ [9].

A plot of preferred orientation (CdS(101)H) peak intensity against cathodic voltage for both as-deposited and CdCl$_2$-treated samples, as depicted in Fig. 5.4, shows that the highest peak intensity was attained at 791 mV under the conditions used in this work. This observation signifies the possibility of growing stoichiometric or near-stoichiometric CdS at 791 mV away from the S-rich CdS obtained at V_g lower than 791 mV or Cd-rich CdS obtained at V_g higher than 791 mV as explained in Sect. 5.3.1. The extracted XRD data from these CdS material work matches the JCPDS reference file No. 01-080-0006 and 00-001-0647 for both the hexagonal and cubic phases, respectively. The crystallite size, D, was calculated using the Scherrer's formula as shown in Equation 3.8. The summary of the XRD data and obtained structural parameters of the CdS (101)H preferred diffraction orientation for layers grown between 780 and 800 mV is shown in Table 5.1, while Table 5.2 summarises all the XRD diffractions observed for CdS thin-film layer grown at $V_g = 791$ mV.

For the CdS layers grown between 780 and 800 nm as shown in Table 5.1, calculated crystallite sizes ranging between 21.9 and 44.0 nm for the as-deposited

Table 5.1 The XRD analysis of CdS layers grown between cathodic potential of 780 and 800 mV for the as-deposited and the CdCl$_2$-treated layers at 400 °C for 20 min in air

Growth voltage (mV)	2θ (°)	Lattice spacing (Å)	FWHM (°)	Crystallite size D (nm)	Plane of orientation (hkl)	Assignments
As-deposited						
780	28.40	3.14	0.389	21.9	(101)	Hex
789	28.47	3.13	0.259	32.9		
790	28.43	3.13	0.259	32.9		
791	28.83	3.09	0.194	44.0		
792	28.39	3.14	0.259	32.9		
793	28.44	3.14	0.259	32.9		
800	28.36	3.14	0.389	21.9		
CdCl$_2$ treated						
780	28.41	3.13	0.389	21.9	(101)	Hex
789	28.34	3.14	0.194	43.9		
790	28.34	3.14	0.194	43.9		
791	28.45	3.13	0.129	65.9		
792	28.42	3.14	0.195	43.9		
793	28.31	3.15	0.195	43.9		
800	28.38	3.14	0.195	43.9		

Table 5.2 The XRD analysis of CdS layers grown at cathodic potential of 791 mV for the as-deposited and the CdCl$_2$ heat-treated layers at 400 °C for 20 min in air

Sample	2θ (°)	Lattice spacing (Å)	FWHM (°)	Crystallite size D (nm)	Plane of orientation (hkl)	Assignments
As-deposited	24.90	–	–	–	(100)	Hex CdS
	26.69	3.34	0.227	37.5	(002)	Hex CdS/FTO
	28.83	3.09	0.195	44.0	(101)	Hex CdS
	38.92	2.31	0.195	45.2	(101)	Hex Cd
	42.64	2.12	0.390	22.9	(319)	Orth S
	47.00	1.93	0.260	34.8	(220)	Cubic CdS
	48.02	1.89	0.195	46.7	(103)	Hex CdS
CdCl$_2$ treated	25.07	3.55	0.260	32.7	(100)	Hex CdS
	26.80	3.32	0.130	65.7	(002)	Hex CdS/FTO
	28.45	3.13	0.130	65.9	(101)	Hex CdS
	48.17	1.88	0.162	56.0	(103)	Hex CdS

and 21.9 and 65.9 nm for the CdCl$_2$-treated CdS layers were observed. Apart from the CdS layer grown at 780 mV, an increase in the crystallite sizes of all the CdS layers was observed after CdCl$_2$ treatment. This observation might be due to the coalescence of crystallites and recrystallisation and reduction in strain/stress

resulting in an overall improvement in the structural properties of CdS layers after CdCl$_2$ treatment. Under the as-deposited and CdCl$_2$ conditions, the highest crystallite sizes were observed at 791 mV.

5.3.3   Raman Study

Figure 5.6 shows typical Raman spectra of ~500-nm-thick layers of both as-deposited and the CdCl$_2$-treated samples of CdS. In this spectra, strong scattering due to the longitudinal optical (LO) vibration mode was observed for both the as-deposited and CdCl$_2$-treated layers. For the as-deposited CdS layer, the dominating 1LO peak was observed at ~303 cm^{-1} while a broad 2LO peak was observed at ~604 cm^{-1}. The 1LO and 2LO peaks for the CdCl$_2$-treated samples were observed at ~300 and 602 cm^{-1}, respectively. The slight red shift observed in the 1LO and 2LO peak positions after heat treatment at 400 °C for 20 min in the presence of CdCl$_2$ in the air is logged in Table 5.3. This shift in Raman peaks can be attributed to internal dislocation and extrinsic defect in CdS layer as a result of tensile or compressive stresses [9]. Tensile and compressive stresses affect the Raman spectrum by a red shift and a blue shift [12]. Furthermore, the obtained FWHM from the Raman spectra

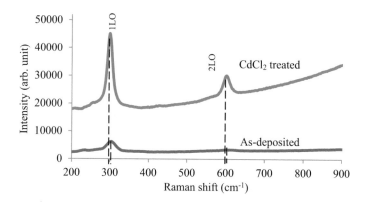

Fig. 5.6   Raman spectra of as-deposited and CdCl$_2$-treated CdS thin films grown at 791 mV

Table 5.3   Raman analysis of as-deposited and CdCl$_2$-treated CdS layers

	As-deposited CdS		CdCl$_2$-treated CdS	
	1LO	2LO	1LO	2LO
Raman peak position (cm^{-1})	303	604	299	601
Intensity (arb. unit)	3645.7	390.6	26,017.0	7504.7
FWHM	28.73	51.44	16.28	20.71

shows a reduction in its value after $CdCl_2$ treatment which is an indication of an improvement in the crystallinity and increased the grain size of the treated layer. This analytical trend is consistent with that observed from the XRD analysis.

5.3.4 Thickness Measurements

The thickness measurement of layers grown between 787 and 793 mV was carried out both theoretically and experimentally. The theoretical estimation was carried out using Faraday's law of electrolysis as discussed in Sect. 3.3 and shown in Equation 3.6. The number of electrons transferred n for the deposition of one molecule of CdS is 2 (i.e. $n = 2$ for CdS). The experimental thickness measurement was carried out using UBM Microfocus optical depth profilometer (UBM, Messtechnik GmbH, Ettlingen, Germany). It was observed in Fig. 5.7a that the value of the measured thickness was lower than the calculated thickness using Faraday's law of electrolysis. This can be attributed to the assumptions made in Faraday's law of electrolysis that all the electronic charges contribute to the deposition of CdS. This assumption is without any consideration of the resistance losses in the system and electronic charge contribution from the decomposition of water into its constituent ions [13].

It was also observed that the growth current density increases with an increase in growth voltage and therefore results in increased film thickness and also changes

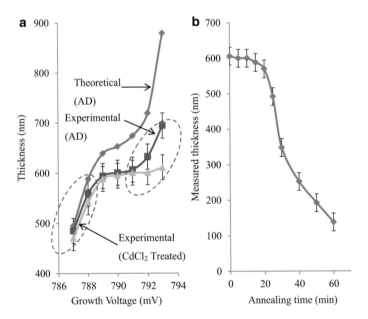

Fig. 5.7 (a) Graph of CdS layer thickness (theoretical and experimental) against growth voltages for both as-deposited and $CdCl_2$-treated CdS layers and (b) graph of measured CdS layer thickness against different annealing duration in air at 400 °C in the presence of $CdCl_2$

were observed in the coloration of the deposited film. It is worth noting that for these experiments, all the CdS layers were grown for the same time duration and heat treated under the same condition. As shown in Fig. 5.7a, after heat treatment at 400 °C for 20 min in the presence of $CdCl_2$, the thickness of the samples grown at 787 and 788 mV show a slight reduction in thickness which might be a result of the loss of sulphur from the S-rich CdS layers. The thickness of the samples grown between 789 and 792 mV remained virtually unchanged which might be attributed to stoichiometric or near-stoichiometric nature of the grown layer within this range. The layers grown above 791 mV show a sharp reduction in thickness as a result of the sublimation of cadmium from the Cd-rich CdS layer. This observation indicates that stoichiometric CdS can withstand the heat treatment, but either S-rich CdS or Cd-rich CdS layer easily break down and sublime under the same conditions.

Further experimentations were performed to determine the effect of heat treatment duration in the presence $CdCl_2$ as shown in Fig. 5.7b. A reduction in thickness due to sublimation of CdS layer was observed after 20 min and a sharp reduction after a 25-min heat treatment duration. Based on this result, heat treatment of CdS in the presence of $CdCl_2$ above 20 min results in layer deterioration and sublimation of the CdS layer and might result into pinhole formation.

5.3.5   Optical Property Analyses

The optical absorbance measurements were carried out on samples grown between 787 and 793 mV for both the as-deposited and the $CdCl_2$-treated samples. This characterisation was performed to determine the energy bandgap of each CdS layer and its conformity with the energy bandgap of the bulk CdS. The spectra of optical absorbance for the CdS thin films grown at different voltages for as-deposited and $CdCl_2$-treated samples are shown in Fig. 5.8a, b. The square of absorbance (A^2) was plotted against the photon energy (hv), and the extrapolated straight-line section of the graph to the photon energy axis at $A^2 = 0$ gives an estimate of the bandgap energy. As observed, the as-deposited CdS layers grown within the explored growth voltage range show bandgaps ranging between 2.41 ± 0.03 eV as shown in Fig. 5.8a and Table 5.4. The $CdCl_2$-treated samples grown within the same growth voltage range show bandgaps between 2.41 ± 0.01eV. This slight reduction in the bandgap range after $CdCl_2$ treatment might be attributed to pinhole removal, recrystallisation of the lattice structure and improvement in material composition. Notably, the growth voltages surrounding 791 mV shows the comparatively sharp difference in the bandgap. This might be due to the insulative property of sulphur due to its richness as observed at low V_g and the metallic behaviour of cadmium due to its richness at high V_g. It was interesting to see that the layer grown at 791 mV shows 2.42 eV bandgap which is comparable with the standard bandgap for bulk CdS.

Also in Table 5.4, it was observed that the growth voltages outside the predetermined/explored range show bandgap values which were not close to the bulk-CdS value of 2.42 eV. This might be attributed to the richness of sulphur or

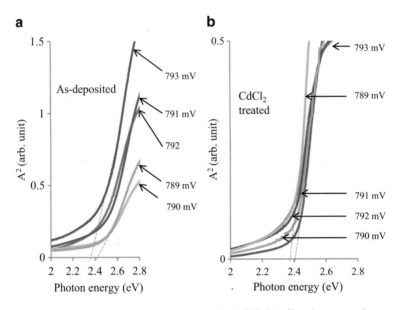

Fig. 5.8 Optical absorption spectra for electrodeposited CdS thin films between voltage range 789 and 793 mV for (**a**) as-deposited and (**b**) CdCl$_2$-treated CdS at 400 °C for 20 min in air

Table 5.4 The optical bandgap and transmittance of CdS layers grown at cathodic potentials between 789 and 793 mV for the as-deposited and the CdCl$_2$-treated layers at 400 °C for 20 min in air

Growth voltage (mV)	789	790	791	792	793
Band gap for as-deposited CdS (eV)	2.45	2.44	2.42	2.42	2.39
Band gap CdCl$_2$-treated CdS (eV)	2.41	2.41	2.42	2.41	2.40
Transmittance for AD-CdS (%)	58	57	47	47	43
Transmittance for CdCl$_2$-treated CdS (%)	79	76	76	63	61

cadmium in the grown CdS layers. Figure 5.9 shows the transmittance of the explored growth voltage range after heat treatment at 400 °C for 20 min in the presence of CdCl$_2$. It is worth mentioning that all the samples show transmittance above 60% for wavelength larger than 512 nm.

5.3.6 Morphological Studies

Figure 5.10a–c show the SEM images of as-deposited (AD) CdS layers grown between 789 and 793 mV, while Fig. 5.10d–f show micrographs of the CdCl$_2$ treated CdS grown between 789 and 793 mV. All the layers were electrodeposited for 2 h on glass/FTO substrates. From observation, all the as-deposited and CdCl$_2$-treated CdS layers explored show full coverage of the underlying glass/FTO

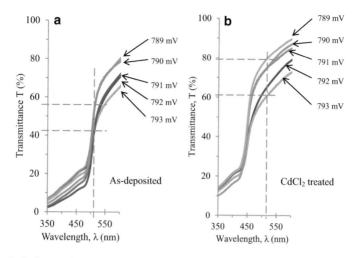

Fig. 5.9 Optical transmittance spectra for electrodeposited CdS thin films between voltage range 789 and 793 mV for (**a**) as-deposited and (**b**) CdCl$_2$ treated at 400 °C for 20 min in air

substrate. For the as-deposited layers, slightly higher grain sizes were observed when grown at 791 mV within the range of 180–225 nm as compared to layers grown at 798 and 793 mV within the range of 130–200 nm. This observation which shows a hint of higher crystallinity at 791 mV might be due to the improvement in the Cd/S stoichiometry towards 1/1 ratio. After CdCl$_2$ treatment, grain growth was observed for all the CdS layers explored. This observation is often associated with improvement in elemental composition, crystallinity, defect passivation, sublimation of excess elemental concentration [14, 15], amongst others.

For the CdS layer grown at 789 mV, the underlying glass/FTO substrates were observed which might be due to the sublimation of excess elemental sulphur. Such gaps/pinholes may serve as shunt paths and reduction in device performance.

5.3.7 Compositional Analysis

EDX measurements were performed to determine the atomic composition of Cd and S for both the as-deposited and CdCl$_2$-treated CdS films. Seven samples were grown at different cathodic voltages within the range of 788–794 mV at 1 mV step and used for the analysis. Figure 5.11 summarises the atomic ratio of Cd to S in CdS layers grown at different growth potentials as observed in EDX. The presence of other elements such as Sn and Cl were also noted. This is due to the underlying glass/FTO layer and the presence of chlorine in the CdS deposition bath and the CdCl$_2$ treatment as described in Sect. 5.2.1.

As illustrated in Fig. 5.11, S-rich CdS layers were observed when grown at voltages below 791 mV (V_i) with atomic ratios of Cd/S less than 1.0. This obtained

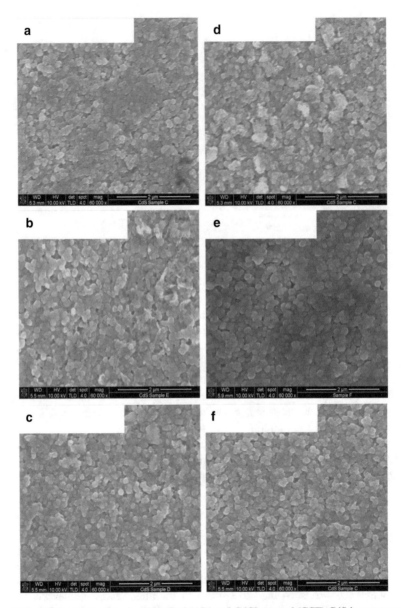

Fig. 5.10 SEM micrographs of as-deposited (AD) and CdCl$_2$-treated (CCT) CdS layers grown at different voltages in the vicinity of 791 mV. (**a**) 789 mV (AD), (**b**) 791 mV (AD), (**c**) 793 mV (AD), (**d**) 789 mV (CCT), (**e**) 791 mV (CCT), (**f**) 793 mV (CCT)

data confirms the S-richness of CdS. A gradual increase in Cd content incorporated into CdS layer was observed with increasing growth voltage. Stoichiometric CdS was observed at 791 mV for both the as-deposited and the CdCl$_2$-treated layers with the Cd/S atomic ratio equal to 1.01. It was interesting to observe that after CdCl$_2$

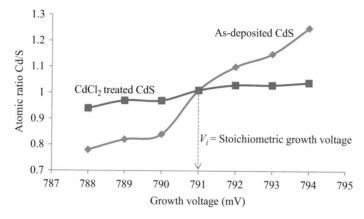

Fig. 5.11 Graphical representation of percentage compositions ratio of Cd to S atoms in as-deposited and CdCl$_2$-treated CdS thin films at different deposition cathodic voltages

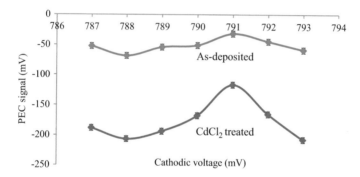

Fig. 5.12 PEC signals for layers grown at different cathodic voltages, showing n-type electrical conductivity type for all samples investigated within explored conditions

treatment, both the S-rich and the Cd-rich CdS tend towards stoichiometry. This observation further shows that CdCl$_2$ treatment improves CdS layer properties by reaction between unreacted Cd and S and the sublimation of excess element from the layer. The result obtained in this section is in agreement with the visual, structural and optical observations earlier discussed in Sects. 5.3.1, 5.3.2 and 5.3.5.

5.3.8   Photoelectrochemical (PEC) Cell Measurement

Figure 5.12 shows the PEC cell measurements of CdS layers grown between the cathodic voltages of (787 and 793) mV under both as-deposited and CdCl$_2$-treated conditions. As observed in Fig. 5.12, both the as-deposited and the CdCl$_2$-treated layers were all n-type in electrical conduction. The n-type conductivity nature of

CdS layers has been reported in the literature as an intrinsic donor defect in CdS layers [16, 17]. This is due to the presence of Cd interstitials and S vacancies in the crystal lattice of the deposited CdS layers.

The magnitudes of the PEC signals in the $CdCl_2$-treated layers were higher than those for the as-deposited layers. This might be as a result of the improvement of the depletion layer at the CdS/electrolyte junction due to the enhancement of the electronic quality of the CdS layer.

5.4   Effect of CdS Thickness

For this investigation, only the $CdCl_2$-treated CdS layers will be discussed due to the triviality of the as-deposited layers. CdS layers grown at 791 mV cathodic voltage with thicknesses ranging from 50 to 200 nm on glass/FTO substrate were explored. The layers were $CdCl_2$ treated for 20 min at 400 °C. The characteristic observations are discussed in the following sections.

5.4.1   X-Ray Diffraction Study Based on CdS Thickness

Figure 5.13a shows the XRD micrographs of glass/FTO/CdS with variation in the CdS thickness between 50 and 200 nm, while Fig. 5.13b shows both the H(101) CdS peak intensity and calculated crystallite size as a function of CdS layer thickness.

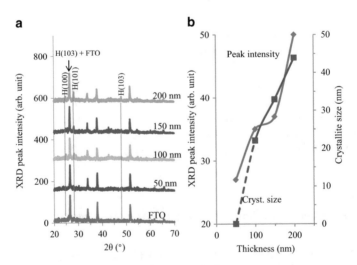

Fig. 5.13 (a) Typical XRD patterns of different CdS layers with increasing thickness after $CdCl_2$ treatment and (b) analysis based on H(101) CdS peak intensity and crystallite size as a function of CdS layer thickness

It should be taken into account that the XRD fingerprints as shown in Fig. 5.13a were stacked together for better peak comparison. From observation, all the layers are polycrystalline in nature with peaks associated with hexagonal (100), (002), (101) and (103) being observed at $2\theta = 24.9°$, 26.7°, 28.4° and 48.0°, respectively, due to the metastable nature of hexagonal CdS [9] after $CdCl_2$ treatment [18].

From observation of the 50-nm-thick CdS layer, it is clear that the peaks associated with H(101) are still at the emerging stage with no visible presence of peak associated with H(100) and H(103). With the increase in CdS thickness to 100 nm and above, peaks attributed to H(100) and H(103) planes were observed along with the H(101). An increase in reflection intensity with increasing thickness was noticeable. Based on this observation, it can be said that the increase in the reflection intensity is directly associated with increasing thickness of the CdS layer [18, 19].

As shown in Fig. 5.13b, the crystallite size was calculated using Scherrer's equation as shown in Equation 3.8. From observation, the H(101) preferred orientation reflection of the 50-nm-thick CdS layer was indistinguishable by the Philips PW 3710 X'Pert diffractometer for analysis due to its thinness or weak crystallinity. It should be noted that the electrodeposition of CdS commences with the deposition of sulphur before the deposition of cadmium is triggered [3]. Therefore, a sulphur-rich CdS and weak CdS can be experienced at the initiation stages of nucleation of CdS. Furthermore, an increase in crystallite size was observed to be associated with the increase of the CdS layer thickness. A similar observation has also been documented in the literature by other independent researchers [18, 19].

5.4.2   Optical Properties Based on CdS Thickness

Figure 5.14 shows the graph of percentage transmittance and the energy bandgap versus wavelength for $CdCl_2$-treated glass/FTO/CdS layer. It was observed that the bandgap lies within 2.44 ± 0.02 eV with the highest bandgap observed at 50 nm. This observation might be due to either the early nucleation stage in which the

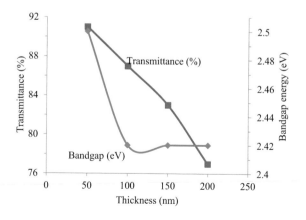

Fig. 5.14 The effect of CdS film thickness on optical bandgap and transmittance spectra for $CdCl_2$-treated CdS thin films at 400 °C for 20 min in air

underlying glass/FTO substrate is not fully covered due to the mechanism of deposition in electroplating as discussed in Sect. 5.4.1 or the comparatively low crystallite size and crystallinity [20] as shown in Fig. 5.13b for the 50-nm-thick CdS layer. Consequently, an increase in the steepness of the absorption edge and a shift of the absorption edge towards longer wavelength were observed with increase in CdS thickness. This observation is well documented in the literature [19, 21] with researchers such as Bosio [15] and Han [22] suggesting superior semiconductor material quality with steeper absorption edge due to lesser impurity energy levels and defects in the thin film.

The reduction in bandgap observed with increase in thickness to 100 nm might be due to the full coverage of the underlying substrate. Consequently, reduction in transmittance from ~90 to ~77% was observed with increasing thickness from 50 to 200 nm as shown in Fig. 5.14. It is important to note that thinner CdS layers give higher transmittance; hence higher photocurrent can be generated. On the other hand, thinner CdS layers have high tendency to suffer from discontinuities and defects such as pinholes [18, 23] due to the nucleation mechanism of the deposition technique. Therefore it is pertinent that the optical property of the semiconductor layer and its thickness must be considered to achieve optimum photocurrent.

5.4.3 SEM Studies Based on CdS Thickness

The SEM micrographs of 50, 100, 150 and 200 nm thicknesses of $CdCl_2$-treated CdS layers grown on glass/FTO are shown in Fig. 5.15a–d, respectively. Although the as-deposited glass/FTO/CdS layers are not presented in this work due to its triviality, grain growth and the coalescence of grains is observed after $CdCl_2$ treatment. As observed in Fig. 5.15a, full coverage of the underlying glass/FTO substrate has not been attained with the glass/FTO grains still visible beneath the CdS layer due to significant gaps between CdS grains. This observation might be as a result of the early stage of nucleation of CdS to the underlying conducting substrate and also the columnar growth mechanism in electroplated semiconductors. With an increase in thickness to 100 nm and above as shown in Fig. 5.15b–d, grains tend to be more closely packed with full coverage of the glass/FTO underlying substrate. It should be noted that the fabrication of a solar cell using the 50 nm pinhole infected CdS layer will lower the performance of the solar cell due to low open-circuit voltage, fill factor and short-circuit current density as a result of shunting.

5.5 Effect of CdS Heat Treatment Temperature

For this set of experiments, ~500 nm thick CdS was electrodeposited at 791 mV on glass/FTO with a dimension (4 × 6) cm^2. After growth, the layer was cut into six pieces of (4 × 1) cm^2. One of the CdS layers was left as-deposited, while the others

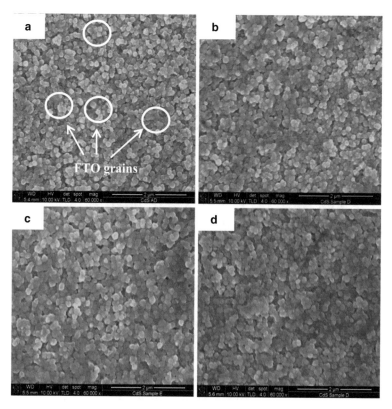

Fig. 5.15 SEM micrograph images for CdCl$_2$-treated CdS with thicknesses of (**a**) 75 nm, (**b**) 100 nm, (**c**) 200 nm, (**d**) 400 nm, respectively

were CdCl$_2$ treated at different temperatures ranging from (380 to 450) °C for 20 min. The observed characteristic properties are discussed below.

5.5.1   X-Ray Diffraction Studies Based on Heat Treatment Temperature

Figure 5.16a shows the XRD diffractions of CdS layers CdCl$_2$ treated at different heat treatment temperature but at constant time duration, while Fig. 5.16b shows the XRD results analysis based on both the H(101) and H(100) CdS diffraction peaks.

As observed in Fig. 5.16a, the preferred orientation for the CdCl$_2$-treated CdS layer between 380 and 400 °C for 20 min is H(101). But between 420 and 440 °C treatment temperature, recrystallisation and competition of H(101) and H(100) phases were observed. Such observation has been documented in the literature on the effect of post-growth treatment temperature on recrystallisation of CdTe materials by Dharmadasa [24].

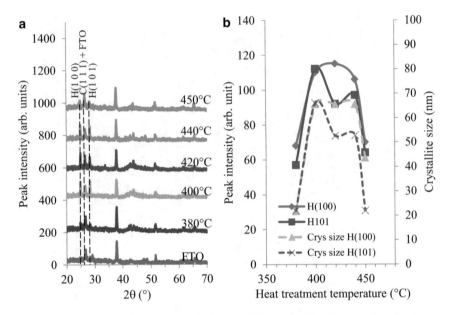

Fig. 5.16 (a) Typical XRD patterns of CdS layers CdCl$_2$ treated for the same time duration at different heat treatment temperatures and (b) analysis based on H(100) and H(101) CdS peaks as a function of CdS thickness

Furthermore, no noticeable reduction in peak intensity was observed after 500 °C treatment temperature, but reduction in crystallite sizes from ~(65 to 52) nm ensued. These observations attest to improvement in crystallinity after CdCl$_2$ treatment as documented in the literature with the optimal treatment temperature at 400 °C due to reduced phase competition and observation of the highest crystallite size of ~65 nm calculated using both the H(101) and H(100) parameters as shown in Fig. 5.16b. Although other parameters need to be optimised, the effect of heat treatment temperature cannot be overlooked.

5.5.2 Optical Properties Based on Heat Treatment Temperature

Figure 5.17 shows the effect of CdCl$_2$ treatment temperature on both optical absorption and transmittance of CdS layers. It is obvious from observation that an improvement in the bandgap is observed at (400–420) °C as compared to the layers treated below 380 and above 420 °C. The slightly high bandgap observed after 380 °C CdCl$_2$ treatment of the CdS layers might be due to the incorporation of excess/unreacted elemental Cd and S resulting into comparatively low crystallinity as also observed in Fig. 5.16. While the increase in the bandgap of CdS, CdCl$_2$ treated above 420 °C might be due to the formation of large pinholes as shown in

Fig. 5.17 The effect of CdCl$_2$ treatment temperature on optical bandgap and transmittance of CdCl$_2$-treated CdS thin films for a constant duration of 20 min in air

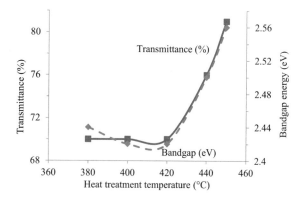

Fig. 5.18 which allows the passage of photons at the explored wavelength and resulting into the high surge in transmittance as shown in Fig. 5.17. At temperatures beyond 420 °C, CdS layer thickness reduces due to sublimation of material and therefore both the measured transmittance and bandgap increases.

5.5.3 SEM Studies Based on Heat Treatment Temperature

Figure 5.18 shows the SEM micrographs of as-deposited and CdCl$_2$-treated CdS layers heat treated for 20 min in air at different temperatures. At constant CdCl$_2$ heat treatment duration for different temperatures, grain growth through the coalescence of smaller grains was observed, with higher CdCl$_2$ treatment temperature correlating with an increase in grain size.

The grain size for the as-deposited CdS was estimated to be between 120 and 200 nm. The CdCl$_2$-treated CdS at 380, 400, 420, 440 and 450 °C for 20 min were estimated to be within the ranges of 120–225 nm, 180–225 nm, 180–300 nm, 300–2000 nm and 300–2000 nm, respectively. At 420 °C (see Fig. 5.18d) cracks were observed, while at 440 °C and above, large pinholes were observed due to intensive coalescence of grains at such high temperatures. Furthermore, this observation might also be due to the mechanism of growth in electrodeposited semiconductor layers. It should be noted that nucleation of electrodeposited semiconductor material on glass/FTO underlying layer commences at the apex of the rough surface of the substrate due to the high electric field experienced at these tips. This mechanism of growth results in the column-like growth leaving pinholes within and around columns; they are gradually filled-up with an increase in layer thickness. Based on observation, layers heat-treated at 400 °C seems better due to higher grain size as compared to the as-deposited CdS layers and CdCl$_2$ treated at 380 °C without cracks and pinholes as observed at 420 °C and above.

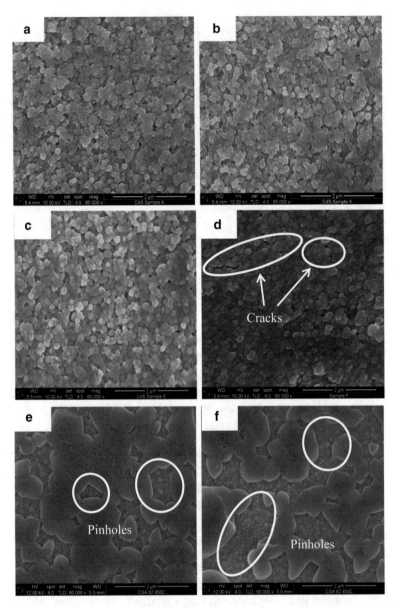

Fig. 5.18 SEM micrograph images of (**a**) as-deposited and CdCl$_2$-treated CdS layers for 20 min in air at (**b**) 380 °C, (**c**) 400 °C, (**d**) 420 °C, (**e**) 440 °C and (**f**) 450 °C, respectively

5.6   Effect of Heat Treatment Duration

These sets of experiments were performed on ~500 nm thick CdS layers which were CdCl$_2$ treated and heated at 400 °C in the air for different time durations.

5.6.1 X-Ray Diffraction Study Based on Heat Treatment Duration

Figure 5.19a shows the XRD diffraction patterns of CdS CdCl$_2$ treated at 400 °C for varied heat treatment durations ranging between 10 and 50 min, while Fig. 5.19b shows the graph of the preferred orientation CdS (H(101)) intensity and the calculated crystallite sizes against CdCl$_2$ heat treatment durations. As shown in Fig. 5.19a, the preferred orientation of all the CdS layer explored were H(101). Based on observation, improvement in the H(101) CdS peak intensity was observed after CdCl$_2$ treatment at 400 °C between 10 and 50 min due to the improvement in both electronic and material properties of the CdS layers as discussed earlier in Sect. 5.3.2 and also recorded in the literature [14, 15].

As shown in Fig. 5.19b, a gradual increase in the H(101) diffraction intensity was observed from the as-deposited layer and peaking at 20-min CdCl$_2$ treatment duration. Above this duration, a gradual reduction in the diffraction intensities was observed, which might be due to CdS layer deterioration through sublimation and increased pinhole density as observed morphologically (see Sect. 5.6.3). For the CdS layers CdCl$_2$ treated for 50 min, a complete collapse of both the H(101) and H(100) diffraction associated with CdS was observed which might be due to high material loss due to sublimation as a result of prolonged heat treatment duration. Furthermore, no improvement in crystallite sizes was observed between the as-deposited CdS layer and the CdCl$_2$-treated layers for the duration of 10–30 min as depicted in Fig. 5.19b. This observation might either be due to the limitation of Scherrer's equation for measuring large crystallites [25] or due to the sensitivity of the XRD equipment utilised in this work. These observations reiterate the importance of CdCl$_2$ treatment duration on crystallinity and crystallite size.

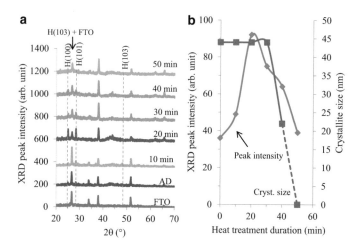

Fig. 5.19 (a) Typical XRD patterns of CdS layers CdCl$_2$ treated and heated at 400 °C for different time durations and (b) graph of the preferred orientation of CdS (H(101)) intensity and the calculated crystallite sizes against CdCl$_2$ heat treatment duration

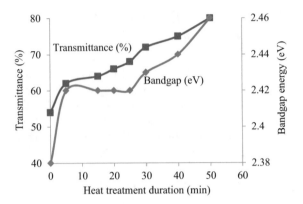

Fig. 5.20 The effect of heat treatment duration on CdS thin film grown at 791 mV and heat treated at different durations at 400 °C in the presence of CdCl$_2$

5.6.2   Optical Properties Based on Heat Treatment Duration

Figure 5.20 shows the effect of heat treatment duration on both the bandgap and transmittance of CdS thin films CdCl$_2$ treated at 400 °C in air. As shown in Fig. 5.20, a gradual increase in transmittance was observed with increasing CdCl$_2$ treatment duration. It should be noted that the change in transmittance between 15 and 25 min treatment duration is comparatively minimal with a transmittance difference of ~65% ± 2% as compared to others. Correspondingly, a similar trend was observed with bandgap energy of the CdS layers as shown in Fig. 5.20. The initial increase in both transmittance and bandgap energy between 0 and 10 min can be attributed to the improvement in CdS properties either by sublimation of excess elements (Cd and S), increase in CdS crystallinity or grain growth. While the CdS layers CdCl$_2$ treated between 15 and 25 min can be said to be highly crystalline CdS layer with full coverage of underlying FTO substrate, evidence of these can be seen in the bandgap which equals the bulk CdS bandgap of 2.42 eV. These properties are required to produce high-efficiency solar cell devices. The increase in both the transmittance and bandgap energy observed for CdS layers CdCl$_2$ treated for durations above 25 min can be attributed to the severe loss of CdS layers and the formation of pinholes. This observation is in accord with the thickness measurement discussed in Sect. 5.3.4.

5.6.3   SEM Studies Based on Heat Treatment Duration

Figure 5.21 shows the SEM micrograph images of (a) as-deposited and CdCl$_2$-treated CdS at 400 °C for the durations of (b) 10, (c) 20, (d) 30, (g) 40 and (h) 50 min, respectively. Topologically, the surface of the FTO/CdS substrate shows full coverage with CdS nanoparticles for the as-deposited and heated samples for 10-, 20- and 30-min duration. It was notable that the coalescence of grains and increase in grain sizes was slightly observable after 10 min heat treatment duration, with grain sizes within the range of 150–200 nm. A further increase in grain size to ~(300–550) nm

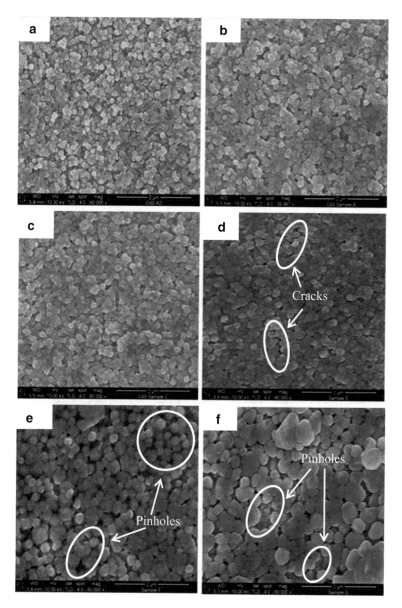

Fig. 5.21 SEM micrograph images of (**a**) as-deposited and CdCl$_2$-treated CdS at 400 °C for the duration of (**b**) 10, (**c**) 20, (**d**) 30, (**e**) 40 and (**f**) 50 min, respectively

was observed with an increase in the heat treatment duration. This microstructural change of the surface morphology as a result of heat treatment can be attributed to recrystallisation and the partial transformation of the mixed cubic/hexagonal phases into single hexagonal phase after heat treatment [26] as observed from XRD.

Gaps along the grain boundaries were observed for samples heat treated for 30 min and above. This can be attributed to the sublimation of the CdS layer and due to a build-up of larger CdS nanoparticles from the coalescence/growth of CdS crystallites [27] in the vertical direction. These observations are in accord with the thickness measurement in Fig. 5.7b.

5.7   Testing the Electronic Quality of CdS

5.7.1   Current-Voltage Characteristics with Ohmic Contacts (DC Conductivity)

The DC conductivity measurements were carried out on the electrodeposited CdS layers to determine the electrical conductivity (σ) and resistivity (ρ) of the layers. These experiments were performed to determine both the effect of film thickness and the heat treatment duration on the electrical properties of CdS layers. Two millimeter diameter and 60-nm-thick indium (In) contacts were evaporated onto CdS layers to give ohmic behaviour. The electrical resistance (R) of the glass/FTO/CdS/In structures were calculated from the ohmic I-V data under dark condition using Rera Solution PV simulation system. Using Equation 3.18, which incorporates resistance R, contact area A and film thickness L, the resistivity (ρ) was calculated.

The summary of the electrical resistivity and conductivity as a function of CdS layer thickness is tabulated in Table 5.5. To determine the effect of heat treatment duration on the electrical properties, 310 nm thick CdS was deposited on a (3×8) cm^2 FTO substrate and treated with CdCl$_2$ after growth. The substrate was cut into (3×1) cm^2 pieces before heat treating each substrate at 400 °C for different time durations. As observed in Fig. 5.22a, an initial stagnation of the resistivity of the

Table 5.5 Summary of electrical properties of CdS layers after heat treatment durations at 400 °C in the presence of CdCl$_2$ in air

Heat treatment duration at 400 °C (min)	Measured thickness (nm)	Avg. resistance (Ω)	Avg. resistivity $\times 10^4$ (Ω cm)	Avg. conductivity $\times 10^{-5}$ (Ω cm)$^{-1}$
5	302	28.4	2.97	3.39
10	300	28.3	2.97	3.39
20	285	20.2	2.23	4.51
25	246	22.8	2.91	3.48
30	174	24.8	4.47	2.32
40	126	20.6	5.13	2.09
50	96	20.4	6.67	1.57
60	69	20.8	9.46	1.09

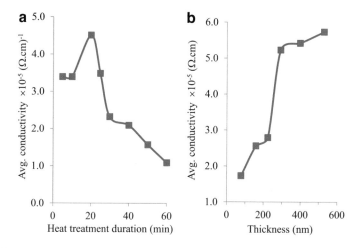

Fig. 5.22 (a) The effect of heat treatment duration on the electrical conductivity of CdS thin films grown at 791 mV and heat treated at different durations at 400 °C in the presence of $CdCl_2$. (b) The effect of CdS film thickness on electrical conductivity for $CdCl_2$-treated CdS thin films at 400 °C for 20 min in air

CdS layer was observed for 5 and 10 min heat treatment duration which suggests that the latent heat required in melting grain boundaries, activate grain growth and forming crystalline mono-phased CdS have not been surpassed due to the gradual temperature increase of the substrate during their treatment time durations. The increase in the conductivity of CdS for samples heat treated for 20 min can be attributed to growth in grain size, reduction of grain boundaries and increased crystallinity of the CdS layer. While the samples heat treated above, 20 min duration shows a gradual decrease in conductivity which might be due to the deterioration of the CdS layer caused by sublimation of elements [3]. This reduction of electrical conductivity could also be due to compensation taking place by diffusion of p-dopants such as Na from glass substrate during prolonged heat treatment. This observation is in accordance with the structural, morphological and optical properties discussed above. This shows that the optimum heat treatment at 400 °C is for ~20 min to achieve the highest electrical conductivity.

Further experimentation on the effect of thickness on the electrical conductivity of CdS layers which had been post-growth treated with $CdCl_2$ at 400 °C for 20 min shows that an increase in film thickness results into an increase in the conductivity as shown in Table 5.6 and Fig. 5.22b. This can be attributed to the formation of large grains, reduction of voids/pinholes and increase in S vacancy and Cd interstitials which serves as n-type intrinsic dopants in CdS [17].

Table 5.6 The effect of layer thickness on the electrical conductivity of CdCl$_2$-treated CdS layers

Thickness (nm)	Resistance (Ω)	Avg. resistivity $\times 10^4$ (Ω cm)	Avg. conductivity $\times 10^{-5}$ (Ω cm)$^{-1}$
79	15.4	6.11	1.73
159	15.4	3.94	2.56
223	25.5	3.58	2.79
292	18.7	2.01	5.23
397	23.5	1.86	5.42
528	30.0	1.79	5.73

5.7.2   Current-Voltage Characteristics with Rectifying Contacts

The current-voltage (I-V) characteristics for rectifying contacts have been broadly used to study Schottky diodes and to determine some important device and material parameters. For these experiments, the thicknesses of CdS films were varied to determine the effect of thickness on the characteristic behaviour of glass/FTO/ CdS/Au Schottky diodes. The I-V characteristics were measured in dark condition at a bias voltage range of −1.00 to 1.00 V. The I-V characteristics of a Schottky diode under dark condition have been previously expressed in Sect. 3.5.1.1.

Figure 5.23a shows a typical semi-logarithmic current vs. voltage curve measured under dark condition for Au Schottky contacts made on heat-treated CdS, Figure 5.23b linear-linear I-V curve of Au Schottky contacts made on heat-treated CdS layers and Fig. 5.23c the effect of heat treatment duration at 400 °C on ideality factor and potential barrier height for Au Schottky contacts under dark condition. The series resistance R_s and shunt resistance R_{sh} were calculated from the slopes of the linear-linear I-V curve in the forward and reverse bias, respectively, as shown in Table 5.7. It was observed that across all thicknesses in this experiment, the cells show infinite (∞) R_{sh} and $R_s \leq 0.1$ Ω which is close to that of an ideal diode with R_{sh} and R_s equal to infinity (∞) and zero (0), respectively. The observed ideality factor was in the range of 1.63–1.83 with a gradual increase in value with an increase in thickness as shown in Table 5.7. This shows that the current transport mechanisms in the depletion region of the M/S structure consist of both thermionic emission and recombination and generation (R&G) processes. Other factors such as the increase in R_s and tunnelling through the diode could have increased the ideality factor n [2], but in this experiment, the R_s was 0.1 Ω. Further observations on the effect of thickness on CdS layer shows that thickness has no significant influence on the barrier height ϕ_b. The fabricated CdS diodes show an RF value of ~10^4 across all thicknesses. This indicates that the CdS layers are suitable for application in electronic devices.

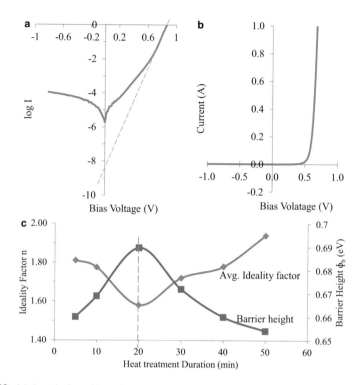

Fig. 5.23 (**a**) A typical semi-logarithmic current vs. voltage curve measured under dark condition for Au Schottky contacts made on heat-treated CdS, (**b**) linear-linear *I-V* curve of Au Schottky contacts made on heat-treated CdS layers and (**c**) the effect of heat treatment duration at 400 °C on ideality factor and potential barrier height for Au Schottky contacts on CdS under dark condition

Table 5.7 The summary of observed parameters obtained from current-voltage (*I-V*) characteristics for CdS/Au Schottky diodes at different thicknesses

Thickness (nm)	R_s (Ω)	R_{sh} (MΩ)	log RF	$I_o \times 10^{-7}$ (A)	n	ϕ_b (eV)
79.0	0.089	0.96	4.2	3.46	1.68–1.72	>0.67
159.6	0.075	1.10	4.2	3.04	1.70–1.72	>0.67
223.7	0.075	1.13	4.1	4.02	1.70–1.75	>0.67
292.4	0.075	1.18	4.0	4.66	1.72–1.78	>0.66
397.5	0.084	1.25	4.2	4.71	1.76–1.82	>0.66
528.0	0.094	1.27	4.1	5.17	1.80–1.83	>0.66

Additional experimentations on the effect of heat treatment duration on CdS layers as depicted in Fig. 5.23c shows an optimal value for both ϕ_b and n at 20 min heat treatment duration. This observation can be attributed to better material quality, high crystallinity and minimisation of defect distribution. Material deterioration was observed at higher heat treatment duration as the n value tends towards 2.00, which show the presence of higher R&G centres.

5.7.3 Capacitance-Voltage Characteristics of Rectifying Contacts

The capacitance-voltage (C-V) technique was performed to determine important device and material characteristics such as position of Fermi level (E_F), built-in potential (V_{bi}), barrier height ϕ_b, doping concentration of the material ($N_D - N_A$), charge carrier mobility ($\mu_\perp$) and depletion layer width at zero bias of glass/FTO/CdS/Au Schottky diodes fabricated on varied CdS layer thickness. The justification for the use of combined I-V and C-V measurements for the estimation of the material's electronic parameters instead of Hall Effect measurement have been discussed in Sect. 3.5.

All the samples used in this experiment were $CdCl_2$ treated at 400 °C for 20 min. The C-V measurements were performed in dark condition at a bias voltage range of −1.00 to 1.00 V with 1 MHz AC signal at 300 K. The built-in potential (V_{bi}) and donor concentration ($N_D - N_A$) in this configuration can be determined using the Mott-Schottky plot as shown in Fig. 5.24 using Equation 3.52.

For the experiments discussed in this book, it has been assumed that all excess donor atoms (N_D) are ionised at room temperature; therefore $n \approx (N_D - N_A)$ as shown in Equation 3.57. The $\mu_\perp$ is the mobility of electrons in the direction perpendicular to FTO surface. It should be noted that $\mu_\perp$ is different from the reported mobility values measured by conventional Hall Effect method. These values are $\mu_{//}$ and it represents the mobility of electrons moving parallel to the FTO layer. Due to the presence of grain boundaries, $\mu_{//}$ will be much smaller than $\mu_\perp$ for nano-crystalline CdS layers.

Further to this, cognisance should be taken of the calculated ϕ_b values using C-V measurements (see Equation 3.58) as it is affected by the effects of defects and interfacial resistive layers producing excess capacitance and inhomogeneity of the semiconductor layer in the diode [28]. The calculated ϕ_b was mainly to show the trend in this work. As observed from Table 5.8 and Fig. 5.25a, the increase in

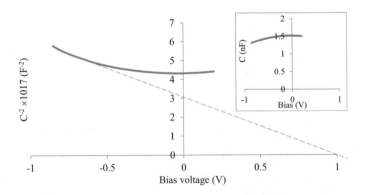

Fig. 5.24 A typical Mott-Schottky plot for Schottky contacts made on electroplated CdS layers. Inset shows the variation of capacitance as a function of DC bias voltage

Table 5.8 The summary of observed parameters obtained from Mott-Schottky plots for CdS/Au Schottky diodes at different thicknesses

Thickness (nm)	Range of ideality factor n	$\sigma \times 10^{-5}$ $(\Omega\ \text{cm})^{-1}$	Avg. $(N_D - N_A) \times 10^{17}$ (cm^{-3})	$(E_C - E_F)$ (eV)	ϕ_b (eV)	$\mu_\perp \times 10^{-4}$ $(\text{cm}^2\ \text{V}^{-1}\ \text{s}^{-1})$
79.0	1.68–1.72	1.73	9.03	0.11	>0.84	1.19
159.6	1.70–1.72	2.56	8.13	0.12	>0.86	1.96
223.7	1.70–1.75	2.79	7.22	0.14	>0.89	2.41
292.4	1.72–1.78	5.23	4.92	0.15	>0.89	6.63
397.5	1.76–1.82	5.42	1.30	0.17	>0.91	26.00
528.0	1.80–1.83	5.73	1.22	0.19	>0.92	29.26

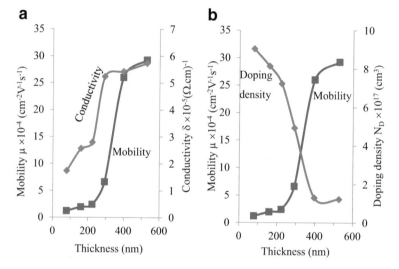

Fig. 5.25 (a) Graph of mobility and conductivity as a function of CdS layer thickness, (b) graph of mobility and doping density as a function of CdS layer thickness

average conductivity with respect to the increasing thickness of CdS layer can be attributed to the resultant effects on n and $\mu_\perp$ for electrons. The gradual increase in n values indicates that R&G centres increase with increase in thickness.

The mobility of charge carriers (free electrons) depends on scattering due to lattice vibration, ionised and neutral impurities, native defects and grain boundaries. Table 5.8 summarises the variation of parameters such as n, σ, $N_D - N_A$ and $\mu_\perp$ for CdS layers with increasing thickness. The diode behaviour observed in Fig. 5.24 is typical for moderately doped semiconductors with the doping range of 10^{15}–10^{17} cm^{-3} [28]. Although there are possibilities of obtaining varying doping concentrations under different growth conditions [28], in the case of this experimental work, all growth parameters were kept constant.

Additional experimentation was performed to determine the effect of heat treatment duration on 205-nm-thick as-deposited CdS layer. The layers were subjected to

Table 5.9 The summary of observed parameters obtained from Mott-Schottky plots and DC conductivity measurements for CdS/Au Schottky diodes and CdS/In ohmic contacts for CdS CdCl$_2$ treated and heated at 400 °C for different time duration

HT duration (min)	$\sigma \times 10^{-5}$ (Ω cm)$^{-1}$	n	$(N_D - N_A) \times 10^{18}$ (cm^{-3})	E_F (eV)	ϕ_b (eV)	$\mu \times 10^{-5}$ (cm^2 V^{-1} s^{-1})
5	2.28	1.80–1.82	5.53	0.15	>0.88	2.57
10	2.24	1.76–1.79	1.71	0.18	>0.91	8.18
20	2.99	1.62–1.67	0.90	0.20	>1.01	20.70
30	2.11	1.70–1.75	5.53	0.15	>0.88	2.38
40	1.79	1.75–1.81	9.03	0.14	>0.82	1.24
50	1.31	1.90–1.97	16.25	0.12	>0.75	0.50

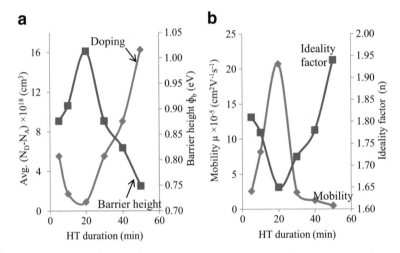

Fig. 5.26 (a) Graphs of barrier height and doping density as a function of heat treatment duration and (b) graphs of mobility and ideality factor as a function of heat treatment duration

post-deposition heat treatment at 400 °C at different heat treatment durations, after which glass/FTO/CdS/Au rectifying contacts were fabricated. As shown in Table 5.9 and Fig. 5.26, low carrier mobility was observed for samples heat treated for a duration less than 10 min. This can be attributed to low CdS crystallinity and resulting high density of grain boundaries and defects as described in Sect. 5.7.2. The mobility reduction for heat treatment duration above 20 min can also be attributed to material deterioration and increase in defects. The heat treatment at 400 °C for 20 min in the air appears to be the optimum condition for post-deposition treatment. Under these conditions, doping concentration of ~9.0 × 10^{17} cm^{-3}, lowest n ~1.65, the highest barrier height of 1.01 eV and the highest mobility $\mu_{\perp}$ ~20.7 cm^2 V^{-1} s^{-1} were achieved for these devices on CdS films.

Table 5.10 The summary of electrical property comparison between ED-CdS, CBD-CdS and bulk CdS crystal

CdS material used	ρ (Ω cm)	RF	n	ϕ_b (eV)	R_s (Ω)	R_{sh} (Ω)	$N_D - N_A$ (cm^{-3})	μ (cm^2 V^{-1} s^{-1})
ED-CdS (this work)	2.2×10^4	10^4	1.65	1.01	0.1	10^6	10^{17}	2.1×10^{-4}
CBD-CdS [28]	10^5	$10^{3.3}$	1.50	1.02	20	–	10^{17}	10^{-3}
Bulk-CdS [29, 30]	10^2	10^4	(1.08–1.30)	1.04 (SXPS) 0.8 (I-V)	–	–	10^{16}	–

To achieve better quality CdS layers required for highly efficient PV solar cell development, the doping density ($N_D - N_A$), electrical conductivity (σ) and charge carrier mobility ($\mu_\perp$) should be optimised with significant consideration of layer thickness, heat treatment temperature and duration. Other factors to consider such as the reduction of defects due to inherent impurities and production of larger grains to reduce native defects and scattering will further improve the properties of the layers. The observation of minimum ideality factor (n) at 20 min heat treatment duration shows the presence of minimum R&G centres under the treatment conditions.

Comparison of results on electrodeposited ED-CdS layers presented in this work with previously reported Schottky barrier work on thin films of CdS grown by chemical bath deposition (CBD) [28] and bulk CdS crystals grown by melt-grown techniques [29, 30] show excellent prospects for development of ED-CdS layers. Table 5.10 summarises the main parameters available for comparison. Schottky barriers formed on all the layers exhibit similar parameters for RF and potential barrier height. Low n values for diodes made on bulk crystal show the presence of less R&G centres as expected. However, the diode properties observed for poly-crystalline CdS show that the layers have good electronic quality for fabricating excellent PV devices. Although the ideality factor n of both ED-CdS and CBD-CdS indicates that the charge carrier transportation mechanism is governed by both thermionic and R&G, it was interesting to observe the comparatively high rectification factor and low series resistance in the ED-CdS layer. The lower barrier height in electrodeposited CdS might be due to Fermi level pinning at the defect states or due to the presence of impurities.

5.8 Conclusions

The work presented in this chapter demonstrates the electrodeposition of CdS using two-electrode configuration from an electrolytic aqueous bath containing cadmium chloride hydrate (CdCl$_2$·xH$_2$O) and thiourea (NH$_2$CSNH$_2$) as cadmium and sulphur precursors, respectively. Based on the material characterisation techniques explored, 791 mV was identified as the best cathodic potential in which stoichiometric ED-CdS is achieved. XRD results show that both cubic and hexagonal CdS were present in the as-deposited CdS layer, while, only the hexagonal CdS was retained after heat treatment at 400 °C for 20 min in the air in the presence of $CdCl_2$ with a preferred orientation along the (002) plane. Both the XRD and SEM results in this work indicate grain growth after $CdCl_2$ treatment and the formation of large clusters of CdS consisting of small crystallites size ranging from ~23 to 47 nm as-deposited and ~33 to 66 nm after $CdCl_2$ treatment and a cluster size ranging from 300 nm to 1 μm. The optical absorption results showed a bandgap range of 2.40–2.42 eV after $CdCl_2$ treatment.

The effects of heat treatment temperature and heat treatment duration were also explored using thickness measurement, SEM, optical absorbance and transmittance. Further analysis of the electronic properties of ED-CdS using both *I-V* and *C-V* measurements were also explored. Excellent rectifying diodes with rectification factor exceeding $\sim 10^4$ were formed with FTO/CdS/Au structures. The observed ideality factor ranges between 1.50 and 2.00 and a barrier height above 1.01 V were achieved. The *C-V* and Mott-Schottky plot of the effect of CdS layer thickness shows an increasing conductivity and hence mobility with an increase in layer thickness. An inverse relationship between doping density and mobility was also observed. Comparisons of Schottky device parameters show that ED-CdS layers have good electronic quality when compared to CBD-CdS layers and bulk CdS crystal. The lowest ideality factor was observed after heat treatment at 400 °C for 20 min which indicates the lowest R&G centres under the condition.

References

1. K.R. Murali, S. Kumaresan, J. Joseph Prince, Characteristics of CdS films brush electrodeposited on low-temperature substrates. Mater. Sci. Semicond. Process. **10**, 56–60 (2007). https://doi.org/10.1016/j.mssp.2006.11.002
2. I.M. Dharmadasa, *Advances in Thin-Film Solar Cells* (Pan Stanford, Singapore, 2013)
3. N.A. Abdul-Manaf, A.R. Weerasinghe, O.K. Echendu, I.M. Dharmadasa, Electro-plating and characterisation of cadmium sulphide thin films using ammonium thiosulphate as the sulphur source. J. Mater. Sci. Mater. Electron. **26**, 2418–2429 (2015). https://doi.org/10.1007/s10854-015-2700-5
4. K. Yamaguchi, P. Mukherjee, T. Yoshida, H. Minoura, Multiple fabrications of crystalline CdS thin films from a single bath by EICD in acidic aqueous solution of Cd2+ and thiourea complex. Chem. Lett. **30**, 864–865 (2001). https://doi.org/10.1246/cl.2001.864
5. M.P.R. Panicker, M. Knaster, F.A. Kroger, Cathodic deposition of CdTe from aqueous electrolytes. J. Electrochem. Soc. **125**, 566 (1978). https://doi.org/10.1149/1.2131499
6. M. Rami, E. Benamar, M. Fahoume, A. Ennaoui, Growth analysis of electrodeposited CdS on ITO coated glass using atomic force microscopy. Phys. Status Solidi A Appl. Res. **172**, 137–147 (1999). https://doi.org/10.1002/(SICI)1521-396X(199903)172:1<137::AID-PSSA137>3.0.CO;2-V

7. A. Izgorodin, O. Winther-Jensen, B. Winther-Jensen, D.R. MacFarlane, CdS thin-film electro-deposition from a phosphonium ionic liquid. Phys. Chem. Chem. Phys. **11**, 8532 (2009). https://doi.org/10.1039/b906995j

8. N.A. Abdul-Manaf, H.I. Salim, M.L. Madugu, O.I. Olusola, I.M. Dharmadasa, Electro-plating and characterisation of CdTe thin films using CdCl2 as the cadmium source. Energies **8**, 10883–10903 (2015). https://doi.org/10.3390/en81010883

9. A. Chandran, K.C. George, Phase instability and defect induced evolution of optical properties in Cd rich-CdS nanoparticles. J. Appl. Phys. **115**, 164309 (2014). https://doi.org/10.1063/1.4873961

10. A. Abdolahzadeh Ziabari, F.E. Ghodsi, Growth, characterization and studying of sol–gel derived CdS nanocrystalline thin films incorporated in polyethyleneglycol: effects of post-heat treatment. Sol. Energy Mater. Sol. Cells **105**, 249–262 (2012). https://doi.org/10.1016/j.solmat.2012.05.014

11. H. Köhler, Optical properties and energy-band structure of CdO. Solid State Commun. **11**, 1687–1690 (1972). https://doi.org/10.1016/0038-1098(72)90772-7

12. A. Tanaka, S. Onari, T. Arai, One phonon Raman scattering of CdS microcrystals embedded in a germanium dioxide glass matrix. J. Phys. Soc. Japan **61**, 4222–4228 (1992). https://doi.org/10.1143/JPSJ.61.4222

13. J. Sun, D.K. Zhong, D.R. Gamelin, Composite photoanodes for photoelectrochemical solar water splitting. Energy Environ. Sci. **3**, 1252 (2010). https://doi.org/10.1039/c0ee00030b

14. I.M. Dharmadasa, Review of the CdCl2 treatment used in CdS/CdTe thin film solar cell development and new evidence towards improved understanding. Coatings **4**, 282–307 (2014). https://doi.org/10.3390/coatings4020282

15. A. Bosio, N. Romeo, S. Mazzamuto, V. Canevari, Polycrystalline CdTe thin films for photo-voltaic applications. Prog. Cryst. Growth Charact. Mater. **52**, 247–279 (2006). https://doi.org/10.1016/j.pcrysgrow.2006.09.001

16. S.D. Sathaye, A.P.B. Sinha, Studies on thin films of cadmium sulphide prepared by a chemical deposition method. Thin Solid Films **37**, 15–23 (1976). https://doi.org/10.1016/0040-6090(76)90531-9

17. C. Wu, J. Jie, L. Wang, Y. Yu, Q. Peng, X. Zhang, J. Cai, H. Guo, D. Wu, Y. Jiang, Chlorine-doped n-type CdS nanowires with enhanced photoconductivity. Nanotechnology **21**, 505203 (2010). https://doi.org/10.1088/0957-4484/22/6/069801

18. A.A. Ojo, I.M. Dharmadasa, Investigation of electronic quality of electrodeposited cadmium sulphide layers from thiourea precursor for use in large area electronics. Mater. Chem. Phys. **180**, 14–28 (2016). https://doi.org/10.1016/j.matchemphys.2016.05.006

19. N.S. Das, P.K. Ghosh, M.K. Mitra, K.K. Chattopadhyay, Effect of film thickness on the energy band gap of nanocrystalline CdS thin films analyzed by spectroscopic ellipsometry. Phys. E Low Dimension. Syst. Nanostruct. **42**, 2097–2102 (2010). https://doi.org/10.1016/j.physe.2010.03.035

20. M.A. Redwan, E.H. Aly, L.I. Soliman, A.A. El-Shazely, H.A. Zayed, Characteristics of n-Cd0.9 Zn0.1S/p-CdTe heterojunctions. Vacuum **69**, 545–555 (2003). https://doi.org/10.1016/S0042-207X(02)00604-8

21. A.K. Mohsin, N. Bidin, Effect of CdS thickness on the optical and structural properties of TiO2/CdS nanocomposite film. Adv. Mater. Res. **1107**, 547–552 (2015). https://doi.org/10.4028/www.scientific.net/AMR.1107.547

22. J. Han, C. Spanheimer, G. Haindl, G. Fu, V. Krishnakumar, J. Schaffner, C. Fan, K. Zhao, A. Klein, W. Jaegermann, Optimized chemical bath deposited CdS layers for the improvement of CdTe solar cells. Sol. Energy Mater. Sol. Cells **95**, 816–820 (2011). https://doi.org/10.1016/j.solmat.2010.10.027

23. M.A. Tashkandi, W.S. Sampath, Eliminating pinholes in CSS deposited CdS films, in *2012 I.E. 38th Photovoltaic Specialist Conference (PVSC 2012)*, pp. 143–146. https://doi.org/10.1109/PVSC.2012.6317587

24. I.M. Dharmadasa, P. Bingham, O.K. Echendu, H.I. Salim, T. Druffel, R. Dharmadasa, G. Sumanasekera, R. Dharmasena, M.B. Dergacheva, K. Mit, K. Urazov, L. Bowen, M. Walls, A. Abbas, Fabrication of CdS/CdTe-based thin film solar cells using an electrochemical technique. Coatings **4**, 380–415 (2014). https://doi.org/10.3390/coatings4030380

25. A. Monshi, Modified Scherrer equation to estimate more accurately nano-crystallite size using XRD. World J. Nano Sci. Eng. **2**, 154–160 (2012). https://doi.org/10.4236/wjnse.2012.23020

26. G. Mustafa, M.R.I. Chowdhury, D.K. Saha, S. Hussain, O. Islam, Annealing effects on the properties of chemically deposited CdS thin films at ambient condition. Dhaka Univ. J. Sci. **60**, 283–288 (2012)

27. A. Taleb, F. Mesguich, T. Onfroy, X. Yanpeng, Design of TiO2/Au nanoparticle films with controlled crack formation and different architectures using a centrifugal strategy, RSC Adv. **5**, 7007–7017 (2015). https://doi.org/10.1039/C4RA13829E

28. N.B. Chaure, S. Bordas, A.P. Samantilleke, S.N. Chaure, J. Haigh, I.M. Dharmadasa, Investigation of electronic quality of chemical bath deposited cadmium sulphide layers used in thin film photovoltaic solar cells. Thin Solid Films **437**, 10–17 (2003). https://doi.org/10.1016/S0040-6090(03)00671-0

29. N.M. Forsyth, I.M. Dharmadasa, Z. Sobiesierski, R.H. Williams, Schottky barriers to CdS and their importance in Schottky barrier theories. Semicond. Sci. Technol. **4**, 57–59 (1989). https://doi.org/10.1088/0268-1242/4/1/011

30. N. Forsyth, I. Dharmadasa, Z. Sobiesierski, R. Williams, An investigation of metal contacts to II–VI compounds: CdTe and CdS. Vacuum **38**, 369–371 (1988). https://doi.org/10.1016/0042-207X(88)90081-4

Chapter 6
CdTe Deposition and Characterisation

6.1 Introduction

Semiconductor materials from the II–VI group have been highly recognised for their importance in providing a range of materials with high optoelectronic conversion efficiency and also because they possess direct energy bandgaps covering the significant portion of the solar spectrum. Cadmium telluride (CdTe) which belongs to the II–VI family is assumed to be one of the most prominent of the group, and its properties have been extensively investigated [1–3]. CdTe has found a wide range of applications as infrared windows, X- and γ-ray detectors and most especially PV solar cells due to its high absorption coefficient of about 10^4 cm^{-1} at 300 K in the visible and near IR regions of the solar spectrum. As documented in the literature, CdTe with a near-ideal direct bandgap of 1.45 eV [4] possesses the ability to absorb 99% of incident photons with energy higher than the CdTe bandgap at 2 μm thickness [5, 6].

The search for low-cost technologies for mass production and high optoelectronic conversion efficiency of photovoltaic solar cells has prompted the use of different thin-film growth techniques in the deposition of CdTe. With over 14 various growth techniques reported in the literature [2], low-cost, simplicity, scalability and manufacturability have been some of the attributes of the electrodeposition technique over others [7]. The literature reveals that $CdSO_4$ has been the main precursor in the electrodeposition of CdTe [1, 8], but other precursors such as $Cd(NO_3)_2$ [9–11] and $CdCl_2$ [12, 13] amongst others are yet to be fully explored. Therefore, this chapter presents the comprehensive details of the growth and characterisation of CdTe deposited using two-electrode electroplating technique from an aqueous solution containing cadmium nitrate.

Furthermore, the effect of extrinsic doping and doping concentration of CdTe with halides (F, Cl and I) from group VII and Ga from group III of the periodic table was explored due to their unique properties and their suitability for use in electronic devices.

© Springer International Publishing AG, part of Springer Nature 2019
A. A. Ojo et al., *Next Generation Multilayer Graded Bandgap Solar Cells*,
https://doi.org/10.1007/978-3-319-96667-0_6

6.2   Electrolytic Bath and Substrate Preparation for CdTe

6.2.1   Electrolytic Bath Preparation

CdTe thin films were cathodically electrodeposited on glass/FTO substrates by the potentiostatic technique. Cadmium nitrate tetrahydrate $Cd(NO_3)_2 \cdot 4H_2O$ of 99.997% purity was used as the cadmium precursor, while tellurium oxide (TeO_2) of 5 N (99.999%) purity was used as the tellurium source. The aqueous electrolyte was prepared with 1.5 M $Cd(NO_3)_2 \cdot 4H_2O$ in 800 mL deionised (DI) water contained in a 1000 mL polypropylene beaker. The polypropylene beaker was placed in an 1800 mL glass beaker containing DI water. The glass beaker serves as the outer water bath and helps to maintain uniform heating of the electrolyte. The Cd $(NO_3)_2 \cdot 4H_2O$ solution was electro-purified for ~50 h to reduce the impurity level. Afterwards, 0.0002 M of TeO_2 was added and stirred for 5 h to achieve homogeneity of the mixed solution prior to electroplating.

It is worth mentioning that, due to the solubility issues of TeO_2 in aqueous solutions (but it is partially soluble in some acidic media [14]), 0.03 M of TeO_2 solution was prepared by dissolving it in 30 mL of concentrated nitric acid (HNO_3) and stirred for about 1 h. Afterwards, the acidic TeO_2 solution was diluted with 400 mL of DI water in a plastic conical flask. An initial 5 mL of the TeO_2 solution was added to the electro-purified aqueous $Cd(NO_3)_2 \cdot 4H_2O$ electrolyte to give a total concentration of 0.0002 M TeO_2. The electrolytic bath containing 1.5 M Cd $(NO_3)_2 \cdot 4H_2O$ plus 0.0002 M TeO_2 in 800 mL of DI water will be referred to as CdTe bath henceforth. Before electrodeposition, the CdTe bath temperature, pH and stirring rate were maintained at ~85 °C, 2.00 ± 0.02 and ~300 rpm, respectively. The pH was adjusted using diluted solutions of HNO_3 and NH_4OH. The CdTe bath setup mimics the 2E configuration as described in Sect. 3.3. The working electrode (or cathode) utilised is glass/FTO with a sheet resistance of 7 Ω/sq. Prior to the deposition of the CdTe layer, the cyclic voltammogram of the resulting CdTe bath was recorded to determine the possible deposition potential range of CdTe.

6.2.2   Substrate Preparation

Substrates were ultrasonically cleaned at the initial stage in soap solution for 20 min and rinsed in deionised (DI) water. The substrates were then cleaned thoroughly with methanol and acetone to remove any grease and rinsed in DI water. Finally, the FTO was submerged in a clean beaker of DI water and transferred directly into the electroplating bath. After deposition, electroplated CdTe layers were rinsed, dried and divided into two halves. One half was left as-deposited, and the other half was cut into several samples and heat treated in the presence of $CdCl_2$ at 400 °C in the air for different time durations to enhance its properties. The $CdCl_2$ treatment was performed by adding few drops of aqueous solution containing 0.1 M $CdCl_2$ in 20 mL of DI water to the surface of the semiconductor layer. The full coverage of the layers with the treatment solutions was achieved by spreading the solution using solution-damped cotton bud. The semiconductor layer was allowed to air-dry and heat treated at 400 °C for 20 min.

6.3   Growth and Voltage Optimisation of CdTe

Aside from the glass/FTO substrate utilised for cyclic voltammetry, the CdTe layers were electrodeposited between the deposition voltages of 1355 and 1385 mV for 3 h. The glass/FTO/CdTe layers were cut into two. Half was left as-deposited, while the other half was CdCl₂ treated as discussed in Sect. 6.2.2. Both the as-deposited and CdCl₂-treated CdTe layers were characterised afterwards for both their material and electronic properties.

6.3.1   Cyclic Voltammetric Study

Figure 6.1 shows the cyclic voltammogram of an aqueous solution of a mixture of 1.5 M Cd(NO₃)₂·4H₂O and 0.0002 M TeO₂ solution in 800 mL of DI water during the forward and reverse cycles between −100 and ~1500 mV cathodic voltage. The scanning rate was fixed at 3 mVs^{-1} and the pH was adjusted to 2.00 ± 0.02. The bath

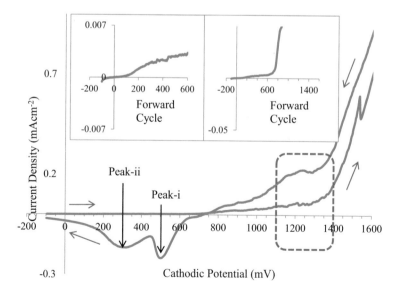

Fig. 6.1 A typical cyclic voltammogram for deposition electrolyte containing the mixture of 1.5 M Cd(NO₃)₂·4H₂O and 0.0002 M TeO₂ solutions at ~85 °C and pH = 2.00 ± 0.02. The insets are the expanded sections of the forward cycle for both Te and Cd deposition initiations

temperature and the stirring speed were kept constant at ~85 °C and 300 rpm, respectively. In the electrodeposition of CdTe, tellurium deposits first, followed by cadmium due to its more positive standard reduction potential (E_o) value of +593 mV as compared to the −403 mV for cadmium [15] with respect to standard H_2 electrode. Ascribable to the high difference between the E_o potential of $HTeO_2^+$ and Cd^{2+}, the concentration of Cd^{2+} is kept much higher than $HTeO_2^+$ in the electrolytic bath [16, 17]. It is observed from Fig. 6.1 that Te starts to deposit in the forward cycle at a cathodic potential of ~130 mV (see in Fig. 6.1) under the experimental conditions used according to the electrochemical reaction as shown in Equation 6.1:

$$HTeO_2^+ + 3H^+ + 4e^- \rightarrow Te + 2H_2O \qquad \text{(Equation 6.1)}$$

As the cathodic potential increases, more Te is deposited, and Cd starts to deposit at ~800 mV. The hump observed at ~800 mV (see inset in Fig. 6.1) depicts the deposition of Cd on the cathode and the co-deposition of CdTe. The rate of deposition of Cd increases with an increase in the cathodic potential. Thus, the electrochemical reaction of Cd deposition and the co-deposition of CdTe are shown in Equations 6.2 and 6.3, respectively:

$$Cd^{2+} + 2e^- \rightarrow Cd \qquad \text{(Equation 6.2)}$$
$$HTeO_2^+ + Cd^{2+} + 3H^+ + 6e^- \rightarrow CdTe + 2H_2O \qquad \text{(Equation 6.3)}$$

In the forward cycle, it was observed that between the cathodic potential range of (1220 and 1400) mV, the deposition current density appears to be stable within the range of 150 and 180 μAcm^{-2}. This range (i.e. 1220–1400 mV) had been pre-characterised at cathodic potential steps of 50 mV using XRD analysis (results are not presented in this report). The highest XRD peak intensity was observed at 1370 mV. Therefore, the surrounding cathodic potential was scanned at steps of 5 mV to identify the best growth potential.

In the reverse cycle, peak-i at ~500 mV represents the dissolution of elemental Cd and Cd from CdTe, while broad peak-ii represents the dissolution of Te at ~309 mV. It is highly imperative that the electrodeposition of CdTe should be done close to the stoichiometric point. Therefore, the cathodic voltage between 1355 and 1385 mV was characterised to determine the optimal growth voltage (V_g) for their material and electronic properties.

6.3.2   X-Ray Diffraction Study

Figure 6.2a, b shows the X-ray diffraction patterns for CdTe layers grown between 1355 and 1385 mV for both as-deposited and $CdCl_2$-treated layers, respectively. Under all the growth voltages and treatment explored, X-ray diffractions associated

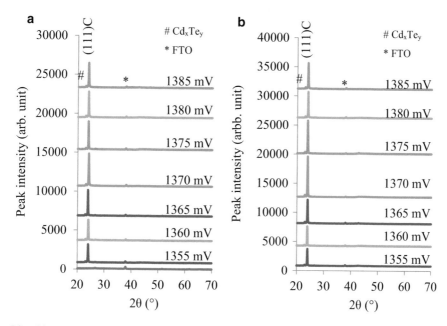

Fig. 6.2 Typical XRD patterns of CdTe layers grown between 1355 and 1385 mV deposition potentials for (**a**) as-deposited CdTe layers and (**b**) CdCl$_2$-treated CdTe layers at 400 °C for 20 min

with Cd$_x$TeO$_y$, predominant (111) cubic CdTe and the underlying FTO were observed at $2\theta = {\sim}21.64°$, 23.90° and 37.62°, respectively.

The incorporated Cd$_x$TeO$_y$ diffraction might be due to the oxidation of the CdTe top layer, while the FTO might either be due to the low thickness of the deposited CdTe layers and the presence of pinholes due to columnar growth mechanism of CdTe layer [11, 18]. It should be noted that the diffractions associated with elemental Te or Cd were not observable due to CdTe growth voltage range within the vicinity of the stoichiometric CdTe.

For both the as-deposited and CdCl$_2$-treated CdTe layers as shown in Figs. 6.2 and 6.3, an increase in the diffraction intensity of the (111) cubic CdTe was observed between 1355 and 1370 mV, while the reduction in the intensity was observed at growth voltages above 1370 mV. This observation signifies that the highest crystallinity and stoichiometric CdTe was attained at 1370 mV, while CdTe layers grown at lower or higher growth voltage suffer from the richness of either Te or Cd, respectively. Furthermore, improvement in the diffraction intensity was observable after CdCl$_2$ treatment of all the CdTe layers grown at different voltages as shown in Fig. 6.3. This observation which is well documented in the literature can be attributed to grain growth and recrystallisation of the crystal structure.

The analysis of the (111) cubic CdTe diffraction for both the as-deposited and CdCl$_2$-treated CdTe layers grown between 1355 and 1385 mV is shown in Table 6.1. From the observation, the highest crystallite size of ~65.3 nm was observed between 1370 and 1375 mV for as-deposited CdTe layers and between 1365 and 1375 mV for

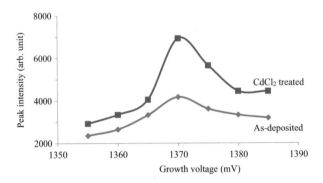

Fig. 6.3 Comparative analysis of CdTe (111)C peak for as-deposited and CdCl$_2$-treated CdTe layers grown at different growth voltages between 1355 and 1385 mV

Table 6.1 The XRD analysis of as-deposited and the CdCl$_2$-treated CdTe layers grown between the growth voltage of 1355 and 1385 mV

Growth voltage (mV)	2θ (°)	Lattice spacing (Å)	FWHM (°)	Crystallite size D (nm)	Assignments
As-deposited					
1355	24.01	3.70	0.195	43.5	(111) cubic
1360	24.04	3.70	0.195	43.5	
1365	23.87	3.71	0.162	52.2	
1370	24.12	3.71	0.130	65.3	
1375	23.97	3.71	0.130	65.3	
1380	23.96	3.71	0.162	52.2	
1385	23.94	3.71	0.162	52.2	
CdCl$_2$ treated					
1355	24.01	3.70	0.162	52.2	(111) cubic
1360	23.95	3.71	0.162	52.2	
1365	23.99	3.71	0.130	65.3	
1370	24.12	3.71	0.130	65.3	
1375	23.97	3.71	0.130	65.3	
1380	23.98	3.71	0.162	52.2	
1385	23.98	3.71	0.162	52.2	

CdCl$_2$-treated CdTe layers. Outside these ranges, lower crystallite sizes were observed, due to the richness of elemental Te or Cd in the CdTe. The extracted XRD data from these CdTe report matches the Joint Committee on Powder Diffraction Standards (JCPDS) reference file No. 01-775-2086. The crystallite size, D, was calculated using the Scherrer's formula (see Equation 3.8).

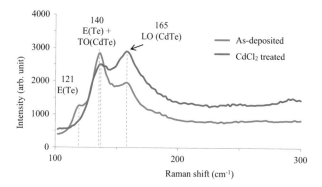

Fig. 6.4 Raman spectra of as-deposited and CdCl$_2$-treated CdTe thin films grown at 1370 mV

6.3.3   Raman Study

Raman spectroscopy studies were performed on samples to determine the phonon modes, identify the phases and determine the crystallinity of both the as-deposited and the CdCl$_2$-treated samples. The excitation source used was a 514 nm argon ion laser. Figure 6.4 shows typical Raman spectra for both as-deposited and the CdCl$_2$-treated CdTe layers grown at 1370 mV. The Raman peak at 165 cm^{-1} corresponds to the longitudinal optical (LO) phonon mode of CdTe, while the peak at 140 cm^{-1} corresponds to the E1 mode of hexagonal Te [19] and fundamental transverse optical (1TO) phonon mode of CdTe. After CdCl$_2$ treatment, the peak intensity observed at 140 cm^{-1} was reduced, and the intensity of the CdTe observed at 165 cm^{-1} was increased due to the formation of CdTe from excess Te and Cd from CdCl$_2$ treatment. Frausto-Reyes (2006) [20] described the increase in the Raman peak intensity as an increase in surface roughness which might be due to grain growth during CdCl$_2$ treatment; it should be noted that due to the combination of both E (Te) and TO(CdTe) at 140 cm^{-1}, the change in intensity cannot be taken as a guide to observe any stoichiometric changes in the layers [21]. Furthermore, the slight Raman shift observed at 164 cm^{-1} can also be attributed to increase in crystallinity [22] after CdCl$_2$ treatment.

While Te diffraction(s) might not be clearly observed in the XRD peak patterns as discussed in Sect. 6.3.2 due to amorphous nature or possible overlapping with FTO peak, a reduction in the E(Te) at 121 cm^{-1} was observable after CdCl$_2$ treatment (see Fig. 6.4). This observation confirms improved crystallinity and stoichiometry after CdCl$_2$ treatment. It should be noted that the presence of elemental Te due to the precipitation of Te in CdTe is peculiar to CdTe grown under any technique [21, 23, 24]. This observation corroborates the summations made on the improvement of crystallinity after CdCl$_2$ treatment in Sect. 6.3.2 [21].

6.3.4 Thickness Measurements

Both theoretically and experimentally, the thickness of electrodeposited CdTe layers was calculated and measured, respectively; the results are shown in Fig. 6.5. The theoretical thickness measurement using Faraday's law of electrolysis serves as the upper limit of thickness due to the assumption that all the electronic charges contribute to the deposition of CdTe without any consideration of the involvement of electronic charges in the electrolysis of water at a voltage above 1230 mV [25]. The mathematical representation of the Faraday's law of electrolysis is shown in Equation 3.6, with predefined parameters. The number of electrons transferred (n) in the deposition of 1 molecule of CdTe is 6 ($n = 6$ for CdTe).

From the summations made from the cyclic voltammetric study in Sect. 6.3.1, it was observed that an increase in deposition voltage results in increase in the deposition current density. The increase in deposition current density can be related to the electrodeposited semiconductor layer thickness (Equation 3.6). Based on this relationship, increase in the thickness of the deposited CdTe layers grown at different cathodic voltages for the same time duration is shown in Fig. 6.5 and can be explained. With the primary emphasis on the experimentally measured thicknesses, it was observed that after $CdCl_2$ treatment, a slight reduction of thicknesses in the CdTe layers measured between 1355 and 1360 mV and 1380 and 1385 was observed, while a considerable retention of thicknesses was observed for CdTe layers grown between 1365 and 1375 mV. These observations might be due to the sublimation of excess elements in the CdTe layers grown outside the growth voltage range in which possible stoichiometric or near-stoichiometric CdTe can be deposited. It should be noted that the melting point of stoichiometric CdTe is ~1092 °C, but the incorporation of excess elements (Cd, Te, O and Cl) serves as impurities which affect the material and electronic properties of the deposited CdTe layers such as the melting temperature. This observation, although not conclusive, iterates the range in which stoichiometric or near-stoichiometric CdTe can be grown.

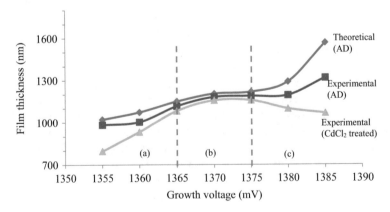

Fig. 6.5 Graphs of thickness against growth voltage under three conditions for layers grown between 1355 and 1385 for the duration of 3 h

6.3.5   Optical Property Analyses

The optical properties of the CdTe thin films were studied at room temperature using Cary 50 Scan UV-Vis spectrophotometer in the wavelength range of 200–1000 nm at room temperature. The measurements were carried out to study the optical absorbance characteristics of CdTe layers grown between the ranges of 1355–1385 mV under both as-deposited and $CdCl_2$ treatments. It should be noted that only the optical absorbance curves of cathodic voltages between 1360 and 1380 mV are shown in Fig. 6.6a, b for better visibility. The values of the obtained bandgaps of the CdTe layers are shown in Table 6.2.

For the as-deposited CdTe layers as shown in Fig. 6.6a and Table 6.2, the range of the bandgap observed spans between 1.42 eV for the lowest growth voltage explored (1355 mV) and 1.50 eV for the highest growth voltage explored (1385 mV). The observation of low bandgap (1.42 eV) might be due to the incorporation of excess Te with a bandgap of 0.34 eV [26]. A narrower bandgap range between 1.48 eV and 1.50 eV was observed for the explored CdTe layers after $CdCl_2$ treatment. This can be attributed to the improvement in the CdTe layers' optical property through the coalescence of smaller grains and grain growth [27, 28], the formation of CdTe from unreacted elements [21], reduction in Te precipitation [21, 23, 29, 30] and change in

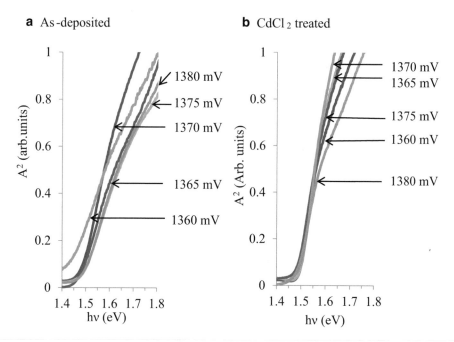

Fig. 6.6 Optical absorption spectra for (**a**) as-deposited and (**b**) $CdCl_2$-treated CdTe thin films grown between 1360 and 1380 mV

Table 6.2 The optical bandgaps of both as-deposited and $CdCl_2$-treated CdTe layers grown between 1355 and 1385 mV

Growth voltage (mV)	As-deposited (eV)	$CdCl_2$ treated (eV)
1355	1.42	1.49
1360	1.43	1.49
1365	1.48	1.49
1370	1.48	1.48
1375	1.48	1.48
1380	1.48	1.48
1385	1.50	1.50

atomic composition [31] amongst other reasons. Further observations based on Fig. 6.6a, b show that the highest gradient of the optical absorption edge occurs at 1370 mV, which indicates the superiority of material quality due to lesser impurity energy levels and defects in the thin film [28, 32].

6.3.6 Morphological Studies

Figure 6.7a–c shows the SEM images of as-grown CdTe layers grown at 1355, 1370 and 1385 mV, respectively, while in Fig. 6.7d–f are the SEM images after $CdCl_2$ treatment, respectively. As shown in Fig. 6.7, all the CdTe layers explored under both as-deposited and $CdCl_2$-treated conditions show full coverage of the underlying substrate. For the as-deposited CdTe layers, the presence of small grains, their agglomeration into curly floral-like grains and high grain boundary density were observed. The cluster sizes vary between ~50 and ~200 nm for CdTe layers grown at 1355 mV, ~200 nm and ~400 nm for CdTe layers grown at 1370 mV and ~50 and ~400 nm for CdTe layers grown at 1385 mV. It can be seen that the intrinsic doping of CdTe with either excess Te or Cd does affect the morphological property of the as-deposited CdTe layers. It should be noted that the incorporation of such high grain boundary density in CdTe layer in the photovoltaic application will promote scattering of charge carriers at the grain boundaries [32, 33] and reduction in charge carrier mobility [34].

After $CdCl_2$ treatment (Fig. 6.7d–f), grain growth, reduction in grain boundary density and full coverage of underlying substrate were observed. Varying grain sizes between ~50 and ~500 nm were observed for CdTe layers grown at 1355 mV, ~200 and ~2000 nm for CdTe layers grown at 1370 mV and ~200 nm and ~1500 nm for CdTe layers grown at 1385 mV.

As documented in the literature, the incorporation of $CdCl_2$ during CdTe heat treatment facilitates grain growth and recrystallization of the CdTe layer [27, 28] coupled with the passivation of grain boundaries, improvement of the Cd/Te atomic stoichiometry and reduction of Te precipitates in CdTe [21]. Furthermore, the $CdCl_2$ treatment of CdTe in an oxygen-containing atmosphere is required to passivate the

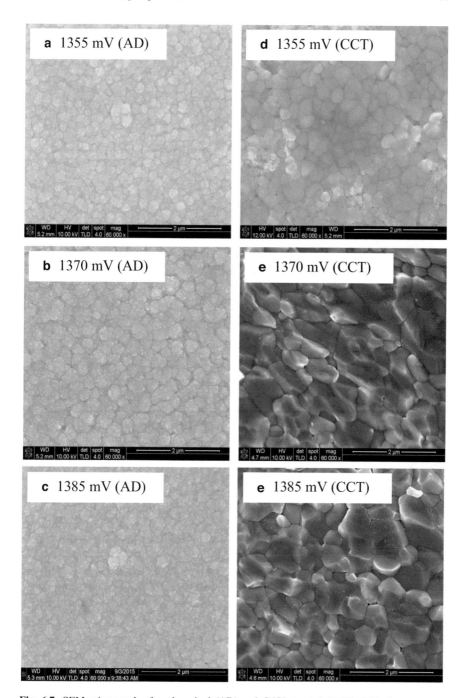

Fig. 6.7 SEM micrograph of as-deposited (AD) and CdCl$_2$-treated (CCT) CdTe layers grown between 1355 and 1385 mV

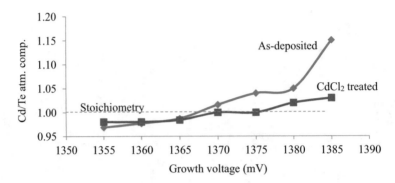

Fig. 6.8 Graphical representation of the composition ratio of Cd/Te for both as-deposited and CdCl$_2$-treated CdTe layers against growth voltage between 1355 and 1385 mV

grain boundaries further and increase carrier concentration [35]. In accordance with the literature, increase in grain sizes was observable at all the explored V_g, with the largest grain size observed from CdTe grown at 1370 mV. This observation further depicts superiority of CdTe layers grown at 1370 mV morphologically.

6.3.7 Compositional Analysis

For better visibility and comparability, Fig. 6.8 shows the graph of the atomic ratio of Cd/Te for all CdTe layers grown between 1355 and 1385 mV under both as-deposited and CdCl$_2$ conditions. Figure 6.9a–c shows the EDX spectra of the as-deposited CdTe layers grown at 1355, 1370 and 1385 mV, respectively, while in Fig. 6.9d–f are the respective EDX spectra after CdCl$_2$ treatment. The unidentified EDX peaks in Fig. 6.9 at 3.5 and 4.5 keV are for Cd and Te, respectively [36]. As shown in Fig. 6.8, CdTe layers grown at voltages lower and higher than 1370 mV show richness in Te and Cd, respectively, under both as-deposited and CdCl$_2$ conditions.

Stoichiometric CdTe layers under both as-deposited and CdCl$_2$-treated conditions were observed at ~1370 mV with the Cd/Te atomic ratio equal to 1.01. After CdCl$_2$ treatment of the CdTe layers, shifts towards stoichiometry of the Cd/Te atomic composition were observed for all the as-deposited CdTe layers. This observation might be due to the dissolution of Te precipitates [21], sublimation of elemental Cd and Te from the CdTe layer and the reaction between unreacted Te with Cd from CdCl$_2$ [27]. This observation is in accord with the summations made in other sections.

6.3.8 Photoelectrochemical (PEC) Cell Measurement

The graphs of PEC measurements of the CdTe thin films grown between 1355 and 1385 mV under both as-deposited and CdCl$_2$-treated conditions against growth voltage are shown in Fig. 6.10. The electrical conductivity type of the CdTe layer

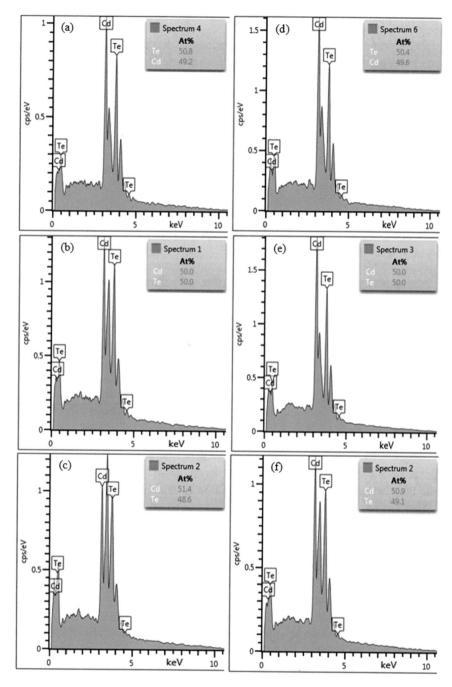

Fig. 6.9 EDX spectra of the (**a**) Te-rich CdTe, (**b**) stoichiometric CdTe and (**c**) Cd-rich CdTe. EDX spectra of CdCl$_2$-treated (**d**) Te-rich CdTe, (**e**) stoichiometric CdTe and (**f**) Cd-rich CdTe. (All as-deposited CdTe layers were grown on the glass/FTO substrates for 2 h duration)

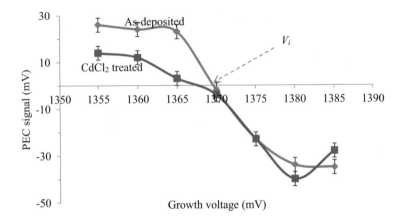

Fig. 6.10 PEC signals for CdTe layers grown at different cathodic potentials, showing a transition from *p*- to *n*-type electrical conduction type at ~1370 mV

is determined by the polarity of the measured voltage difference as described in Sect. 3.4.7. As shown in Fig. 6.10, electrodeposited CdTe layers can be deposited as either an *n*-, *i*- or *p*-type conductivity without extrinsic doping of CdTe layers [37]. From observation, the as-deposited CdTe layers deposited below 1370 mV were *p*-type, while the layers deposited at 1370 mV and above were *n*-type in electrical conduction. A transition growth voltage (V_i) from *p*-type to *n*-type CdTe was observed in-between 1365 and 1370 mV. After $CdCl_2$ treatment, a shift in the atomic composition ratio towards stoichiometry was observed. This might be due to the reaction between unreacted Te with Cd from $CdCl_2$ and the sublimation of excess elemental Cd and Te from the layer. The V_i depicts the growth voltage in which the atomic ratio of Cd to Te is at 1:1.

Figure 6.11 shows the comparison between the compositional analysis as discussed in Sect. 6.3.7, structural analysis as discussed in Sect. 6.3.2 and the PEC cell measurement for the as-deposited CdTe layers grown between 1355 and 1385 mV.

It can be deduced from Fig. 6.11 that the conductivity type of the as-deposited CdTe layers electroplated under the conditions of this work is intrinsically affected by the elemental composition of Cd and Te. Te richness gives *p*-CdTe and Cd richness results in *n*-CdTe. The highest crystallinity of CdTe can be achieved at V_i (~1368 mV). After $CdCl_2$ treatment, although a shift towards the opposing conduction types was observed for both the *p*- and the *n*-type CdTe layers, their conductivity types were retained. The observed shift towards stoichiometry can be attributed to the sublimation of excess element and the formation of CdTe via a chemical reaction between Cd from $CdCl_2$ and excess Te. Aside from the above-mentioned, it should be noted that conductivity-type change after $CdCl_2$ treatment may also be due to the heat treatment temperature, duration of treatment, initial atomic composition of Cd and Te, the concentration of $CdCl_2$ utilised in treatment, defect structure present in the starting material and the material's initial conductivity type as documented in the literature [11, 12, 23, 27].

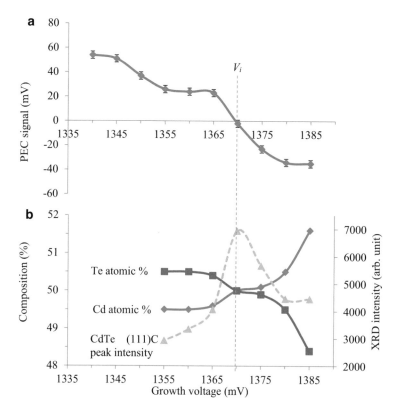

Fig. 6.11 (**a**) PEC signals and (**b**) (111) XRD peak intensity for cubic CdTe and atomic composition of CdTe layers grown at different cathodic potentials between 1340 and 1385 mV, from electrolyte containing $Cd(NO_3)_2$ and TeO_2 solution. The p- to n-transition and the highest crystallinity of CdTe occur at the growth voltage of V_i when the composition changes from Te-rich to Cd-rich (Adapted from Ref [38])

6.4 Effect of CdTe Thickness

For this set of experiments, $CdCl_2$-treated CdTe layers of thicknesses ranging between ~600 and ~1800 nm were utilised. The CdTe layers were grown at a pre-optimised growth voltage of 1370 mV as discussed in Sect. 6.3. It should be noted that the effects of thicknesses of the as-deposited CdTe layers were not explored due to its triviality.

6.4.1 X-Ray Diffraction Study Based on CdTe Thickness

Figure 6.12a shows the XRD patterns of CdTe layers with different thicknesses, while Fig. 6.12b shows the graph of the (111)C CdTe peak intensity and crystallite size as a function of CdTe layer thicknesses.

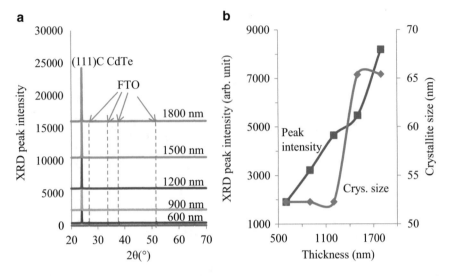

Fig. 6.12 (a) Typical XRD patterns of different CdTe layer thicknesses after CdCl₂ treatment and (**b**) analysis based on (111)C CdTe peak intensity and crystallite size as a function of CdTe thickness

For all the explored CdTe thicknesses, the preferred orientation of (111) cubic CdTe was retained. Furthermore, a direct relationship between CdTe layer thickness and (111)C CdTe diffraction intensity was also observable as shown in Fig. 6.12b, with the increase in CdTe layer thickness resulting in increase in (111)C CdTe intensity. With an emphasis on the crystallite size/CdTe thickness relationship as shown in Fig. 6.12b, an increase in the crystallite size as calculated using Scherrer's equation was observed. As explained in the literature, the increase of the crystallite size (due to a decrease in FWHM) with an increase in CdTe layer thickness reflects the decrease in the concentration of lattice imperfections due to the decrease in the internal micro-strain within the films and an increase in the crystallite size [39].

6.5   Testing the Electronic Quality of CdTe

6.5.1   Current-Voltage Characteristics with Ohmic Contacts (DC Conductivity)

For this study, CdCl₂-treated CdTe layers deposited at different growth voltages between 1355 and 1385 mV were utilised. The thicknesses of the CdTe layers utilised were maintained at ~1000 nm. Ohmic contacts were formed with the CdTe layers depending on the conductivity type. One hundred-nanometer-thick gold (Au) contacts were evaporated on *p*-CdTe layers to give glass/FTO/*p*-CdTe/Au configuration, while 100-nm-thick indium (In) contacts were evaporated on the

Table 6.3 Table of resistance, resistivity and conductivity as a function of growth voltage

Growth voltage (mV)	Conduction type	Resistance R (Ω)	Resistivity $\rho \times 10^3$ (Ω cm)	Conductivity $\sigma \times 10^{-4}$ (Ω cm)$^{-1}$
1355	p	34.84	7.10	1.41
1360	p	22.44	7.05	1.42
1365	p	22.65	7.12	1.41
1370	i	28.08	8.82	1.13
1375	n	16.37	5.14	1.94
1380	n	11.16	3.51	2.85
1385	n	5.28	1.66	6.03

n-CdTe to give glass/FTO/n-CdTe/In configuration. The metallisation processes were performed at 10^{-5} N m^{-2}.

Table 6.3 shows the resistance (R), resistivity (ρ) and conductivity (σ) as a function of CdTe layer growth voltage. The resistances R were derived from the I–V curves generated under the condition as described in Sect. 3.4.8 and using associated Equations 3.19 and 3.20. As shown in Table 6.3, a gradual increase in the conductivity with increase in the cathodic voltage was observed due to the incorporation of more Cd. It should be noted that the increase in the resistance for the CdTe layer grown at 1370 mV is due to the possible 1/1 atomic composition of Cd/Te in CdTe.

6.6   Extrinsic Doping of CdTe

Based on electronic and optoelectronic properties, CdTe has proved to be an excellent II–VI semiconductor material with a near-ideal direct bandgap of 1.45 eV [4] at ambient temperature for a single p-n junction. CdTe has attracted increasing interest due to its application in nuclear radiation detectors [40] and also in photovoltaic (PV) application. With photovoltaics (PV) being within the confines of this book, CdTe-based solar cells have been well explored with its main deficiency being the formation of tellurium precipitates distributed randomly over the whole volume of the CdTe layer during growth [3, 24, 41]. The formation of Te precipitates within the CdTe layer creates uncertainty in the post-growth and post-growth treatment stoichiometry control. Post-growth treatment in the presence of Cd [42] and halogens such as Cl, F and I [43–46] has been documented in the literature to reduce the Te precipitation [21]. Fernandez (2003) [24] has also demonstrated the possibility of eliminating Te precipitation from the bulk of CdTe wafer using gallium melt treatment, although high Te precipitation density was observable on the opposite side of the surface not in contact with the Ga melt treatment. This observation was described as the capability of Ga to dissolve the Te precipitates in CdTe [24, 42].

Although it has been argued that the complete elimination of Te precipitate in CdTe through growth modification or post-growth treatment conditions is impossible [24], this work focuses on the possibility of eliminating Te precipitates by the inclusion of different dopants such as Cl, F, I and Ga (in different CdTe baths) at different concentrations during the electrodeposition of CdTe. Furthermore, this section also evaluates the effect of these dopants on the electronic properties of CdTe to facilitate the device construction and thus promote their practical applications.

6.6.1   Effect of F-Doping on the Material Properties of CdTe

CdTe has been extensively explored and tailored towards photovoltaic applications, especially the CdS/CdTe thin-film configuration which has achieved efficiency up to 21.5% in 2015 [47, 48]. The efficiency of the fabricated cells depends on all the layers and the corresponding interface states with special emphasis on the post-growth treatment with halogens such as chlorine and fluorine. $CdCl_2$ post-growth treatment has been reported in the literature as an essential and sensitive step due to the enhancement in the photoelectrical, structural and morphological properties of CdTe layers which are crucial to achieving high-efficiency solar cells [23, 27, 28]. With an emphasis on fluorine, Rios-Flores [49] and Echendu et al. [50] amongst other authors have also reported further enhancement in CdTe properties with the inclusion of fluorine during CdTe growth and $CdCl_2$post-growth treatment, respectively. The incorporation of fluorine is deemed to increase the conductivity of the CdTe layer due to the excess electrons supplied by fluorine. The literature on the optimal concentration of fluorine in CdTe has not been well established in electrodeposited CdTe. Therefore, this work focuses on the optimisation of extrinsic fluorine-doping concentration in CdTe thin-film electrodeposition bath to achieve better material and device properties.

6.6.1.1   Growth of F-Doped CdTe Layers

Further to the CdTe bath preparation as described in Sect. 6.2.1, cadmium fluoride (CdF_2) of analytical reagent grade of 5N (99.999%) with a varied concentration between 0 and 50 ppm was incorporated into different electrolytic baths with the same measured precursors. The electrolyte was prepared by dissolving 1 M Cd $(NO_3)_2 \cdot 4H_2O$ in 300 mL of deionised water contained in a 500 mL plastic beaker. Other bath conditions are a replica of the CdTe bath described in Sect. 6.2.1. All the CdTe:F layers deposited from all the F-doped CdTe baths were deposited at the pre-optimised cathodic voltage of 1370 mV based on the optical, structural, morphological and compositional analysis (see Sect. 6.2.1). A constant thickness of ~1100 nm was maintained for all the CdTe:F layers.

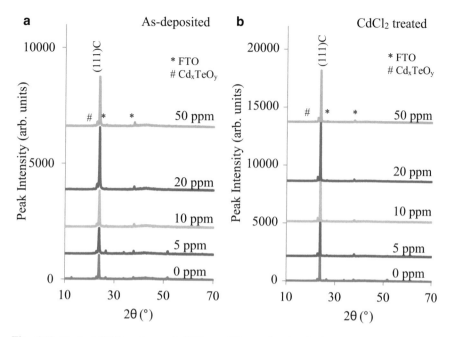

Fig. 6.13 Typical XRD patterns of CdTe at different fluorine-doping concentrations for (**a**) as-deposited CdTe layers and (**b**) CdCl$_2$-treated CdTe layers at 400 °C for 20 min in air

6.6.1.2   Material Properties of CdTe:F

Structural Analysis

The structural properties of the undoped and fluorine-doped CdTe layers are presented in Fig. 6.13. The CdTe layers deposited are polycrystalline with the significant presence of FTO reflections marked (*); Cd$_x$TeO$_y$ reflection was observed at $2\theta = 23.01°$ and prominent (111) cubic CdTe at $2\theta = 23.90°$, while no reflection is associated with fluorine or fluorine compounds due to its presence in low concentration [51]. An increase in the intensity of the preferred CdTe (111)C phase was observed for the as-deposited and CdCl$_2$-treated CdTe layers with increasing F-doping concentration as shown in Fig. 6.13. The XRD intensity reflection appears to have saturated at ~20 ppm for both as-deposited and CdCl$_2$-treated layers with a steady reduction in intensity above 20 ppm doping (see Fig. 6.14). This observation suggests the solubility limit of fluorine in the CdTe lattice is ~20 ppm, although further experimentation is still required. Increasing the F-doping concentration above this point results in a reduction in CdTe (111)C reflection intensity, incomplete crystallisation and low adhesion to the underlying FTO substrate [52, 53] as shown in Fig. 6.14. This observation can be explained by the initial replacement of tellurium ions with fluorine ions in the CdTe lattice up to 20 ppm which aids the crystallisation of CdTe as shown in Fig. 6.13. However, a

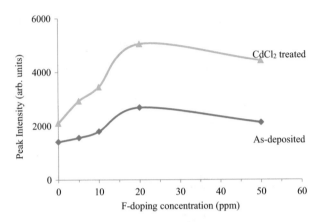

Fig. 6.14 Comparative analysis of CdTe (111)C reflection intensity for as-deposited and CdCl₂-treated layers grown at different fluorine-doping concentrations

further increase in F-doping concentration may readily be incorporated within the crystal lattice but not occupying the proper lattice position whereby increasing disorderliness results within the crystal structure [54, 55]. Further to this, there is a possible formation of highly acidic hydrofluoric (HF) acid which may contribute to continuous etching of the glass/FTO and CdTe:F surfaces [5] and attacking Cd to give Te-rich CdTe layer [56].

The XRD reflection intensity of layers improved after CdCl₂ post-growth treatment at 400 °C for 20 min in the air as shown in Figs. 6.13b and 6.14. This might be attributed to grain growth and recrystallisation of the crystal structure. The highest intensity was observed at 20 ppm fluorine concentration. For clarity, the observed CdTe (111)C reflection intensity was plotted against the concentration of F-doping in CdTe bath as shown in Fig. 6.14. The initial reductions in full width at half maximum (FWHM) and resulting increase in crystallite size were observed with increasing F-doping concentration to 20 ppm in the as-deposited CdTe layers as shown in Table 6.4, while a further increase in F-doping concentration resulted in a reduction in the crystallite size. This observation can be attributed to the replacement of Te atoms from group VI (atomic radius of 1.4 Å) with F atoms having a comparatively low atomic radius of ~0.5 Å from group VII. This might lead to the contraction of the CdTe lattice, hence, a reduction in crystallite size [57, 58] with increasing F atom concentration as observed in the as-deposited CdTe layer grown above 20 ppm concentration as shown in Table 6.4. The calculated crystallite sizes, D, for all the CdCl₂-treated CdTe layers using Scherrer's equation, were all ~52.4 nm. The observation of the same value (~52.4 nm) for all F-doping concentration indicates the limitation of the use of Scherrer's equation for materials with larger grains [59].

The extracted XRD data from this CdTe work matches the Joint Committee on Powder Diffraction Standards (JCPDS) reference file number 01-075-2086-cubic.

Table 6.4 The XRD analysis of CdTe layers grown at different fluorine concentrations

F concentration (ppm)	2θ (°)	Lattice spacing (Å)	FWHM (°)	Crystallite size D (nm)	Plane of orientation (h k l)	Assignments
As-deposited						
0	23.90	3.72	0.260	32.6	(111)	Cubic
5	23.91	3.72	0.259	32.8	(111)	Cubic
10	23.88	3.73	0.227	37.4	(111)	Cubic
20	23.90	3.72	0.227	37.4	(111)	Cubic
50	23.91	3.72	0.292	29.1	(111)	Cubic
CdCl$_2$ treated						
0	23.90	3.72	0.162	52.4	(111)	Cubic
5	23.93	3.72	0.162	52.4	(111)	Cubic
10	23.90	3.89	0.162	52.4	(111)	Cubic
20	23.90	3.74	0.162	52.4	(111)	Cubic
50	23.89	3.72	0.162	52.4	(111)	Cubic

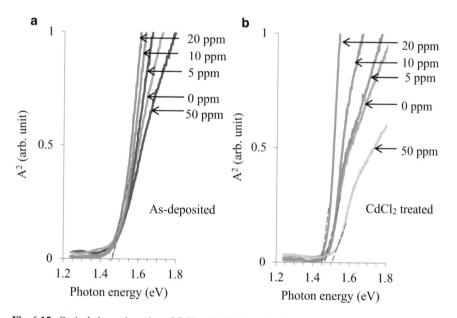

Fig. 6.15 Optical absorption edge of CdTe with different doping concentrations of fluorine for (**a**) as-deposited and (**b**) CdCl$_2$-treated (at 400 °C for 20 min in air) layers

Analysis of Optical Absorption

The optical bandgap of the as-deposited and CdCl$_2$-treated CdTe layers was analysed by plotting the A^2 against photon energy ($h\nu$) as shown in Fig. 6.15. Noticeably in Fig. 6.15a, stronger absorption edges were observed with increasing

F-doping concentration in the as-deposited CdTe layers with no observable influence on the bandgap energy. The observed bandgap energy for the test samples falls within the range of 1.47–1.50 eV. A dissimilar trend was observed after CdCl$_2$ treatment of the CdTe layers as shown in Fig. 6.15b. For CdTe doped with 0–20 ppm fluorine, a rise in the absorption edge, coupled with a reduction in the bandgap energy, was observed. These observations can be attributed to improvement in material properties such as grain size and crystallinity amongst others. An increase in the doping above 20 ppm resulted in a reduction in the gradient of the optical absorption edge and also an increase in the optical bandgap [28, 60]. This might be due to material quality deterioration such as sublimation of CdTe layer. It could be deduced from the optical parameters that fluorine-doping concentration of ~20 ppm gives the best optical property with 1.46 eV bandgap. This observation is in accord with the structural and morphological observations.

Morphological Analysis

Figure 6.16a–c represents the as-deposited CdTe layers containing 0, 20 and 50 ppm F-doping concentration, respectively, while in Fig. 6.16d–f are the resulting micrographs after CdCl$_2$ treatment, respectively. The as-deposited layers show full coverage of the underlying FTO substrate with a noticeable reduction in grain size and indistinct crystal boundary at 50 ppm F-doping concentration. The morphology

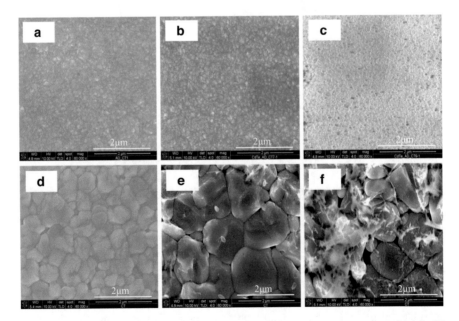

Fig. 6.16 SEM micrographs for CdTe layers grown with F-doping concentration of (**a**) 0, (**b**) 20 and (**c**) 50 ppm and after CdCl$_2$ treatment of CdTe with F-doping of (**d**) 0, (**e**) 20 and (**f**) 50 ppm

reveals that the formation of grain boundaries has been suppressed due to the presence of superfluous fluorine at high concentration. The CdCl$_2$-treated layers show grain growth of ~1, 1.9 and 1.4 µm for 0, 20 and 50 ppm F-doping concentrations, respectively. It could be said that the addition of fluorine to ~20 ppm is advantageous to grain growth, while the inclusion of fluorine to 50 ppm is detrimental as observed in Fig. 6.16f. Although large grains were observed, the formation of pinholes and loss of CdTe layer will be detrimental to device parameters due to shunting.

Compositional Analysis

The presence of base elements (Cd and Te) was observed at all doping concentrations of fluorine, and their percentage concentration for the as-deposited CdTe thin films is presented in Table 6.5. It was observed that the atomic composition of Cd was greater than Te in all the explored fluorine-doping concentration range with an increase in the Cd/Te atomic composition ratio as shown in Table 6.5. This observation may not be due to the continuous replacement of Te ions with F ions at ppm level in the crystal lattice. It is important to note that due to the presence of nominal fluorine in ppm level in the electrolytic bath, the atomic concentration of fluorine in CdTe thin films at different concentrations cannot be determined. However, the presence of F in the electrolyte seems to act as a catalyst to produce Cd-rich layers. As argued by Dharmadasa et al., the defect levels in Cd-rich material are lower than those in the Te-rich material [61]. Therefore, the CdTe layers doped with F should produce solar cell devices with enhanced parameters.

In addition to the presence of Cd and Te atoms, the EDX results may show the presence of Sn, Si, F and O which is due to the underlying glass/FTO substrate and the oxidation of CdTe film surface. The observation made from the compositional analysis gives a clue why the conductivity type is always n-type.

Table 6.5 Summary of the compositional analysis of as-deposited CdTe layers at different F-doping concentration in CdTe bath

F-doping concentration (ppm)	Atomic composition (%)		Cd/Te ratio
	Cd	Te	
0	51.70	48.30	1.07
5	51.80	48.20	1.07
10	52.20	47.80	1.09
20	53.50	46.50	1.15
30	53.40	44.60	1.19
50	55.30	44.70	1.25

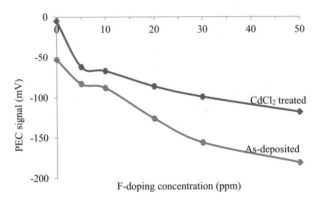

Fig. 6.17 Photoelectrochemical cell measurement for both as-deposited and CdCl$_2$-treated fluorine-doped CdTe thin films

Photoelectrochemical (PEC) Cell Study

Figure 6.17 shows the typical photoelectrochemical cell measurement results for CdTe doped with different concentrations of fluorine. It was observed that both the as-deposited and CdCl$_2$-treated CdTe layers were all *n*-type in electrical conduction. This might be due to the presence of fluorine, an *n*-type dopant in the electrolytic bath. It has been well documented in the literature that the inclusion of halogens in CdTe serves as electron donor impurities by replacing Te atoms in the CdTe lattice with halogen atoms with higher valence electrons. This introduces additional free electrons and therefore makes the CdTe *n*-type. It should be noted that *n*-, *i*- and *p*-CdTe layers can be grown from the 0 ppm F-doping (i.e. undoped) CdTe bath through growth voltage alteration [11].

As shown in Sect. 6.6.1.2 (see compositional analysis), the presence of F in the bath replaces and reduces the excess deposition of Te. The Cd richness makes the material more *n*-type in electrical conduction. The increase in observed PEC signal with increasing F-doping concentration might be due to increasing free electrons from the F-doping and increased Cd richness.

As observed in Fig. 6.17, the PEC signals for all CdCl$_2$-treated CdTe layers in this work tend to reduce. This difference can be attributed to the improvement in the material's electronic properties by the sublimation of excess element during the annealing process or the formation of CdTe through a chemical reaction between precipitated Te in the layer and excess Cd from CdCl$_2$. Furthermore, the observed changes can also be attributed to the doping effect of fluorine.

DC Conductivity Study

The DC conductivity measurements are carried out on glass/FTO/*n*-CdTe/Al structures with different F-doping concentrations. The CdTe layers utilised in this

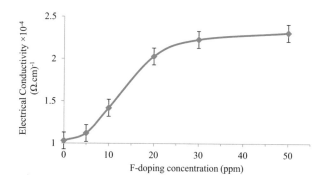

Fig. 6.18 Graph of electrical conductivity against F-doping concentration in CdTe bath

Table 6.6 Electrical resistivity and conductivity as a function of fluorine-doping concentration in the electrolytic bath

F-doping concentration (ppm)	Resistance (Ω)	Resistivity $\times 10^3$ (Ω cm)	Conductivity $\times 10^{-4}$ $(\Omega \, cm)^{-1}$
0	18.75	9.72	1.03
5	17.15	8.97	1.12
10	15.29	5.91	1.73
20	10.52	5.51	2.03
50	8.60	4.60	2.23

experiment were ~1.1 μm thick, and CdCl$_2$-treated samples were heated at 400 °C for 20 min in air. The evaporated Al on the CdTe was to make ohmic contacts, while the I–V measurements were taken in dark conditions. By measuring several ohmic contacts, the average resistance of the glass/FTO/n-CdTe/Al structures was determined. The electrical resistivity and conductivity were calculated with known contact area and the film thickness.

A typical DC conductivity against F-doping concentration in the CdTe bath is plotted in Fig. 6.18. It was observed that the introduction of fluorine increases the electrical conductivity of the layers which saturates at ~20 ppm fluorine concentration (Table 6.6). The saturation of DC conductivity further suggests the solubility limit of fluorine in CdTe lattice [54] as discussed in Sect. 6.6.1.2 (see structural and morphological analyses). Therefore, an increase in F-doping above 20 ppm results into superfluous addition under this experimental condition. As explained in Sect. 6.6.1, the inclusion of fluorine into the crystal lattice introduces free electrons which result in increased conductivity. The enrichment of Cd also contributes to the electrical conductivity, and the resultant effect is measured using this method.

Summations

In this work, we have explored the effect of fluorine doping in a CdTe electrolytic bath on the CdTe layer as it affects it's structural, optical, morphological and compositional properties. An optimal F-doping concentration of 20 ppm in the electrolytic bath was observed under all material characterisations explored in this work due to the solubility limit achieved at 20 ppm F-doping and deterioration in material property afterwards.

6.6.2   Effect of Cl-Doping on the Material Properties of CdTe

Based on the initial observation of the improvement in CdTe material property with increasing concentration of chlorine doping in the CdTe bath, CdTe was grown from an electrolytic bath with the chlorine-based precursor. Therefore, the effect of growth voltage on the characteristic properties of the CdTe:Cl layers will be explored.

6.6.2.1   Growth of Cl-Doped CdTe Layers

CdTe:Cl thin films were electrodeposited cathodically on glass/FTO substrates by a potentiostatic technique in which the anode was a high-purity graphite rod (see Sect. 3.3). The main difference to the CdTe bath as described in Sect. 6.3 is the utilisation of 1.5 M cadmium chloride hydrate $CdCl_2 \cdot xH_2O$ of 99.995% purity as cadmium precursor, while the setup and substrate preparation remained identical.

6.6.2.2   Material Properties of CdTe:Cl

Cyclic Voltammetric Study

Figure 6.19 shows the cyclic voltammogram of an aqueous solution containing a mixture of 1.5 M $CdCl_2 \cdot xH_2O$ and a low level of preprepared TeO_2 solution in 400 mL of DI water during the forward and reverse cycle between 100 and -2000 mV cathodic voltage. The electrolytic bath pH was adjusted to 2.00 ± 0.02 using dilute solutions of HCl and NH_4OH. The stirring rate and bath temperature were maintained at 300 rpm and ~85 °C, respectively. Tellurium deposits first in the electrodeposition process due to its standard reduction potential value of $+593$ mV with respect to the standard H_2 electrode which is more positive than that of cadmium with standard reduction potential value of -403 mV.

It was observed from Fig. 6.19 that the deposition of Te starts in the forward cycle at the cathodic potential of ~400 mV under the experimental conditions used (inset in Fig. 6.19) according to the electrochemical reaction similar to Equation 6.1.

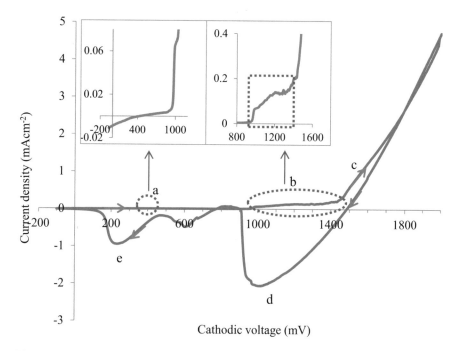

Fig. 6.19 A typical cyclic voltammogram for deposition electrolyte containing the mixture of 1.5 M $CdCl_2 \cdot xH_2O$ and 0.0002 M TeO_2 solution at ~85 °C and pH = 2.00 ± 0.02. The scan rate was set to 3 mVs^{-1}. The insets are the expanded sections of the forward cycle for both Te and Cd deposition initiations

With an increase in cathodic potential, more Te is deposited. The deposition of Cd and formation of CdTe start around ~1000 mV cathodic potential as depicted by the first hump shown in Fig. 6.19. Tellurium-rich CdTe is expected at the initial stage of deposition; afterwards, with an increase in deposition potential, there exists a narrow cathodic potential window in which stoichiometric CdTe compound can be deposited. Further increase in the cathodic potential above this point increases the cadmium richness of electrodeposited CdTe. The electrochemical equation for Cd deposition and the complete CdTe formation are similar to Equations 6.2 and 6.3, respectively.

It should be noted that the steep increase in current density (marked c) at a cathodic voltage above ~1400 mV is due to either the formation of Cd dendrites on the working electrode or the electrolysis of water at any deposition potential above ~1230 mV [14]. Both scenarios have been reported to have a detrimental effect on the quality of the deposited CdTe layer [62]. However, the release of most active hydrogen atoms on the material surface also introduces an advantage of hydrogen passivation during growth although the formation of hydrogen bubbles could be a disadvantage due to a possible cause of material layer peeling. Therefore, selecting the growth voltage has a crucial effect on the growth of a suitable CdTe layer. In the reverse cycle, the dissolution of elemental Cd and Cd from CdTe occurs in the voltage range of ~1500 and ~1000 mV, while the dissolution of Te from the glass/FTO substrate occurs below 750 mV as shown in Fig. 6.19. Therefore, based on the information obtained from the voltammetric study, a cathodic potential range between 1000 and 1400 mV was pre-characterised using XRD for as-deposited CdTe layers grown at a step size of 50 mV (results not presented in this book). The highest peak intensity signalling highest crystallinity was observed at 1350 mV; therefore, surrounding cathodic potentials were scanned at 10 mV intervals and characterised to identify the best cathodic potential in which stoichiometric and near-stoichiometric CdTe can be achieved.

X-Ray Diffraction Study

This study aims to identify the cathodic potential in which stoichiometric or near-stoichiometric CdTe can be grown. This was done by observing the level of crystallinity through XRD peak intensity. Figure 6.20a, b shows typical X-ray diffraction intensity plotted against 2θ angle for layers grown between 1330 and 1400 mV for both as-deposited and $CdCl_2$-treated CdTe layers. For both the as-deposited and $CdCl_2$-treated CdTe layers, only cubic CdTe phases were observed. CdTe (111)C peak corresponding to $2\theta = $ ~23.8° is the dominant XRD peak and the preferred orientation of the electrodeposited CdTe at all growth voltages and conditions. Peaks attributed to CdTe (220)C and (311)C corresponding to $2\theta = $ ~38.6°

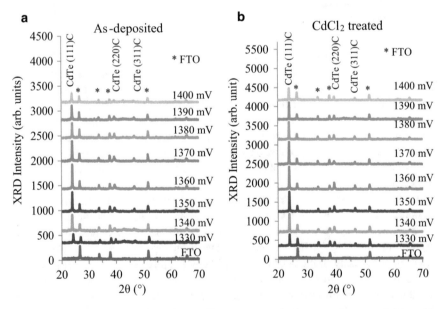

Fig. 6.20 Typical XRD patterns of CdTe layers grown between 1330 and 1400 mV deposition potential for (**a**) as-deposited CdTe layers and (**b**) $CdCl_2$-treated CdTe layers at 400 °C for 20 min in air

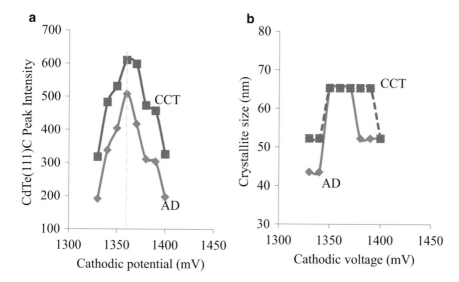

Fig. 6.21 (a) Typical plots of CdTe (111) cubic peak intensity against cathodic voltage. (b) A typical plot of the crystallite sizes against cathodic voltage for the as-deposited and CdCl₂-treated CdTe layers grown between 1330 and 1400 mV

and $2\theta = \sim45.8°$ were also observed aside from the FTO peaks observed at $2\theta = 20.6°$, $33.8°$, $37.9°$, $51.6°$, $60.7°$ and $65.6°$.

For better comparison, XRD patterns were shifted up in the graph as shown in Fig. 6.20a, b. As observed in both Fig. 6.20a, b, the highest XRD peak intensity of the CdTe (111)C for both as-deposited and CdCl₂-treated CdTe layers was observed at 1360 mV. This suggests that highly crystalline CdTe corresponding to stoichiometric CdTe can be electrodeposited at 1360 mV cathodic voltage.

Figure 6.21a shows the comparison of CdTe (111)C peak intensity as a function of the cathodic potential at which the layers were grown, while Fig. 6.21b shows the graph of crystallite size against cathodic voltage.

It is clear from Fig. 6.21a that the treatment of the CdTe layer with CdCl₂ improved the level of crystallinity of the CdTe layers grown at all the explored cathodic voltages. This improvement can be attributed to grain growth and recrystallisation. Furthermore, this improvement could also arise from the formation of CdTe from unreacted excess Te presented in the layer reacting with Cd from CdCl₂ treatment. Full details of this recent understanding are reported by Dharmadasa et al., 2016 [21]. When the growth potential deviates from V_i, the crystallinity suffers due to the presence of two phases: CdTe and Te at lower cathodic potential and CdTe and Cd at high cathodic voltage.

The extracted XRD data from these CdTe work matches the JCPDS reference file No. 01-775-2086. The crystallite size, D, was calculated using the Scherrer's formula (see Equation 3.8). The summary of XRD data and obtained structural parameters of CdTe thin films grown at cathodic voltages between 1330 and 1400 mV using (111) cubic peak are tabulated in Table 6.7.

Table 6.7 The summary of XRD analysis for CdTe layers grown for 2 h at cathodic potentials between 1330 and 1400 mV for the as-deposited and the CdCl$_2$-treated layers at 400 °C for 20 min in air

Cathodic voltage (mV)	2θ (°)	Lattice spacing (Å)	FWHM (°)	Crystallite size D (nm)
As-deposited				
1330	24.23	3.67	0.195	43.6
1340	23.85	3.73	0.195	43.6
1350	23.94	3.71	0.130	65.3
1360	24.02	3.70	0.130	65.4
1370	23.97	3.71	0.130	65.4
1380	23.99	3.71	0.162	52.3
1390	23.93	3.72	0.162	52.3
1400	23.95	3.71	0.162	52.3
CdCl$_2$ treated				
1330	23.95	3.71	0.162	52.3
1340	23.92	3.72	0.162	52.3
1350	23.94	3.71	0.130	65.3
1360	23.95	3.71	0.130	65.3
1370	23.91	3.72	0.130	65.3
1380	23.91	3.72	0.130	65.3
1390	23.97	3.71	0.130	65.3
1400	23.92	3.72	0.162	52.3

As observed in Table 6.7 and Fig. 6.21b, for the as-deposited CdTe layers, low crystallite size was observed at cathodic voltages of ±20 mV away from 1360 mV due to either Cd or Te richness of CdTe grown at these cathodic voltages. After CdCl$_2$ treatment, improvement in crystallite size was observed for all layers. This observation is in accordance with the effect of CdCl$_2$ treatment on CdTe as reported in the literature [27, 28]. However, it should be noted that there is a limitation of the use of Scherrer's equation in determining the crystallite size. This equation is formalised to calculate smaller grains [59] of polycrystalline materials and may not be suitable for the highly crystalline material. Therefore, as shown in Table 6.7 and Fig. 6.21b, the crystallite size saturates at ~65 nm. This must be due to the limitation of Scherrer's equation and the XRD measurement system.

Thickness Measurements

Figure 6.22 shows the graph of electrodeposited CdTe layer thickness estimated using both experimental and theoretical methods against cathodic voltage for CdTe layers grown for 120 min duration. The thickness of the layers grown between 1330 and 1400 mV was calculated theoretically using Faraday's law of electrolysis as shown in Equation 3.6.

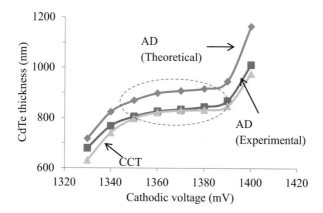

Fig. 6.22 Graph of CdTe layer thickness (theoretical and experimental) against cathodic voltages for both as-deposited and CdCl$_2$-treated CdTe layers

As observed in Fig. 6.22, the value of the calculated thickness using Faraday's law of electrolysis was higher than the measured thickness. It should be noted that Faraday's law of electrolysis assumes that all the electronic charges flowing through the electrolyte contribute to the deposition of the CdTe layers without considering the electronic charges involved in the decomposition of water into its constituent ions. It was further observed that an increase in the cathodic voltage results in an increase in current density and hence affects deposited CdTe layer thickness. After CdCl$_2$ treatment, a slight reduction in thickness was observed. This might be due to the sublimation of CdTe or excess elemental Cd or Te. This also can be due to the formation of a denser layer after CdCl$_2$ treatment. Between 1350 and 1390 mV cathodic voltages, observation of a slightly uniform and almost constant thickness is an indication of the formation of nearly stoichiometric CdTe layers with close deposition current density.

Optical Absorption Study

Using the data acquired through Carry 50 Scan UV-Vis spectrophotometer, the square of the absorbance (A^2) was plotted against the photon energy (hv) as shown in Fig. 6.23a, b for as-deposited and CdCl$_2$-treated CdTe layers grown at different growth voltages, respectively. The straight line segments were extrapolated to $A^2 = 0$ axis to estimate the energy bandgaps of the CdTe layers tabulated in Table 6.8 for comparison.

As observed in Table 6.8, the bandgap of both the as-deposited and the CdCl$_2$-treated CdTe layers within the explored cathodic voltage range falls within 1.45 ± 0.01 eV which is comparable with the bandgap of bulk CdTe [32]. As shown in Fig. 6.24, the growth of CdTe at 1360 mV shows the sharpest absorption edge which signifies superior CdTe layer [28, 32] due to lesser impurity energy

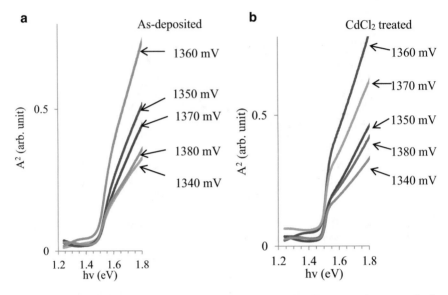

Fig. 6.23 Optical absorption spectra for electrodeposited CdTe thin films grown between cathodic voltage range between 1330 and 1400 mV; (**a**) for as-deposited and (**b**) for CdCl$_2$-treated CdTe at 400 °C for 20 min in air

Table 6.8 The optical bandgap and slope of absorption edge of CdTe layers grown at cathodic voltages between 1330 and 1400 mV for the as-deposited and the CdCl$_2$-treated layers at 400 °C for 20 min in air

	Bandgap (eV)		Slope of absorption edge (eV^{-1})	
Cathodic voltage (mV)	As-deposited	CdCl$_2$ treated	As-deposited	CdCl$_2$ treated
1340	1.46	1.46	1.61	3.13
1350	1.46	1.46	2.13	3.13
1360	1.45	1.45	3.03	5.55
1370	1.45	1.45	1.75	4.35
1380	1.46	1.46	1.75	2.94

levels and defects in the thin film, while the growth of CdTe layer away from 1360 mV shows a reduction in the slope of the optical absorption edge in both the as-deposited and CdCl$_2$-treated CdTe layers due to Cd or Te richness in CdTe. It should be noted that CdCl$_2$ treatment utilised in this work increases the sharpness of the absorption edge across all growth voltages explored. This further attests to the improvement in the optical absorption of CdTe layer after CdCl$_2$ treatment as reported in the literature.

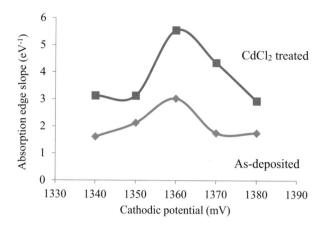

Fig. 6.24 Graph of absorption edge slope against cathodic potential for as-deposited and CdCl$_2$-treated CdTe thin films

Morphological and Compositional Analysis

For both the morphological and compositional experiments, CdTe layers were grown at cathodic voltages between 1330 and 1400 mV on glass/FTO for 120 min. Each glass/FTO/CdTe layer was divided into two halves; one half was left as-deposited, while the other was CdCl$_2$ treated. Figure 6.25 shows the SEM images of as-deposited and CdCl$_2$-treated CdTe grown at 1330, 1360 and 1400 mV.

From topological observation, all the as-deposited CdTe layers show full coverage of the glass/FTO layer. The as-deposited CdTe layers show clear agglomeration of small crystallites forming into cauliflower-like clusters. After CdCl$_2$ treatment, larger crystals were formed through recrystallization and coalescence of grains [62]. The presence of gaps was observable after CdCl$_2$ treatment in all the layers. For CdCl$_2$-treated CdTe layer grown at 1330 mV as shown in Fig. 6.25d, high density of pinholes was observed. This might be due to the Te richness, as a result of the deviation from stoichiometric value and properties as explained by Dharmadasa et al. [62]. The Cd-rich CdTe layer grown at 1400 mV shows less detrimental effect aside from few pinholes along the grain boundaries. The major problem with the presence of pinhole in between grain boundaries is shunting due to contact between the back metal contact and the underlying substrate in solar cell structures. In comparison, larger grains are observed after CdCl$_2$ treatment for the layer grown at 1360 mV. The layers grown at 1400 mV show better coverage of the underlying substrate, although the observed grains were smaller as compared to the layers grown at 1360 mV.

It should be noted that based on new material and device understanding in CdS/CdTe-based solar cells, the richness of Cd in CdTe layer has been demonstrated as more advantageous in high-efficiency device fabrication [63–66] as compared to Te-rich CdTe absorber layer.

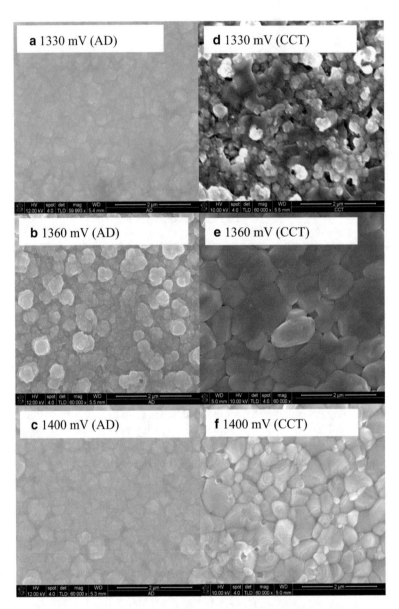

Fig. 6.25 SEM micrographs for CdTe layers grown at 1330, 1360 and 1400 mV (**a–c**) for as-deposited and (**d–f**) for CdCl$_2$-treated layers at 400 °C for 20 min in air

Figure 6.26 shows the compositional analysis of as-deposited and CdCl$_2$-treated CdTe layers using EDX. As observed in Fig. 6.26, both the as-deposited and CdCl$_2$-treated CdTe layers display Te richness at a cathodic voltage lower than 1360 mV, while CdTe layers grown at 1360 and above are rich in Cd.

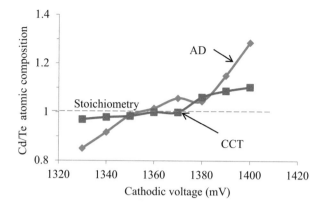

Fig. 6.26 Graphical representation of Cd/Te atomic composition in as-deposited and CdCl$_2$-treated CdTe thin films at different deposition cathodic voltages

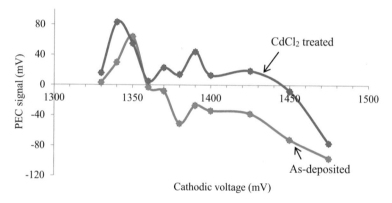

Fig. 6.27 PEC signals for layers grown at different cathodic voltages between 1330 and 1480 mV for both as-deposited and CdCl$_2$-treated CdTe layers

Stoichiometric CdTe was observed at ~1360 mV for both the as-deposited and the CdCl$_2$-treated layers with the Cd/Te atomic ratio equal to 1.01. After CdCl$_2$ treatment, a shift in the atomic composition ratio towards stoichiometry was observed. This might be due to the reaction between unreacted Te with Cd from CdCl$_2$ and the sublimation of excess elemental Cd and Te from the layer. This result is in accord with the observations in Sect. 3.2.6.

Photoelectrochemical (PEC) Cell Study

Figure 6.27 shows the PEC cell measurement results for CdTe layers grown at cathodic voltages between 1330 and 1480 mV in both the as-deposited and CdCl$_2$-treated CdTe samples. Prior to the commencement of this experiment, the PEC cell was calibrated using a known *n*-type CdS layer. For the as-deposited CdTe layers shown in Fig. 6.27, cathodic voltages lower than 1360 mV show *p*-type electrical conductivity, while layers grown at 1360 mV cathodic voltage and above were *n*-type in electrical conduction. This is due to the Te richness in the CdTe layers at lower growth voltages and Cd richness in CdTe grown at higher cathodic voltages. This observation can be related to the redox potential of both Cd and Te and also on the cyclic voltammetric study as explained in Sect. 3.1. It can be deduced that stoichiometric or near-stoichiometric CdTe can be achieved at the cathodic voltage between *n*- and *p*-CdTe layers where the atomic ratio of Cd to Te is at 50:50. This observation is in line with the high crystallinity level observed at 1360 mV in Sect. 3.2.1. The presence of only one phase (CdTe) at this voltage increases the crystallinity of the layer.

After $CdCl_2$ treatment, a shift in the PEC signal towards the *p*-type region was observed across the cathodic range explored. It should be noted that conductivity-type change after $CdCl_2$ treatment could depend on factors such as the heat treatment temperature, duration of treatment, initial atomic composition of Cd and Te, the concentration of $CdCl_2$ utilised in treatment, defect structure present in the starting material and the material's initial conductivity type as documented in the literature [11, 12, 23, 27]. It was interesting to see that CdTe layers grown at 1450 mV and above still retain their initial *n*-type conductivity after $CdCl_2$ treatment. This is due to the effect of more Cd inclusion during deposition.

This observation is crucial to moving towards an understanding of $CdCl_2$ treatment. It is clear that composition change is one of the factors determining the electrical conductivity of CdTe layers. Te richness produces *p*-CdTe, while Cd richness produces *n*-CdTe. The addition of Cl into CdTe is a complex issue. Substitution of Cl into Te sites makes the material *n*-type by acting like a shallow donor [3]. However, there is experimental evidence for the formation of a defect level at 1.39 eV below the conduction band during $CdCl_2$ treatment [29]. This shows that during the $CdCl_2$ treatment, acceptor-like defects are also formed closer to the top of the valence band and this leads to the formation of *p*-type doping of CdTe. One explanation could be that Cl forms a complex with a currently unknown native defect. Therefore, Cl seems to act as an amphoteric dopant in CdTe.

In addition to the above doping effect, self-compensation can take place during heat treatment in the presence of $CdCl_2$ due to the existence of numerous native defects in the material. Therefore, the final electrical conductivity depends on the most dominant process taking place during this treatment.

The experimental evidence in Fig. 6.27 shows the real situation: (1) in as-deposited layers, Te richness produces *p*-CdTe, and Cd richness produces *n*-CdTe; (2) $CdCl_2$ treatment tends to change the material from *n*-properties to *p*-properties. In other words, the Fermi level (FL) moves from the upper half to the lower half of the bandgap. The tendency of the FL crossing the midpoint depends on the initial nature of the material layer. If the Cd richness is dominant in the layers, the FL remains in the upper half of the bandgap keeping the material *n*-type in electrical conductivity. In this discussion, effects of external impurities have been neglected. If an external impurity with dominant doping is introduced in the CdTe layer, the above analogy can be changed.

DC Resistivity

The DC resistivity experiment was performed on CdTe layers grown between the cathodic voltage of 1330 and 1400 mV. The CdTe layers were grown for 120 min each. Two-millimeter diameter of 100-nm-thick gold contacts were evaporated at a low pressure of 10^{-5} Nm^{-2} on the p-type CdTe layer for both the as-deposited and CdCl$_2$-treated CdTe layers (glass/FTO/p-CdTe/Au), while 100-nm-thick indium contacts were evaporated at a low pressure of 10^{-5} Nm^{-2} on the n-CdTe layers for both the as-deposited and CdCl$_2$-treated CdTe layers (glass/FTO/n-CdTe/In) to achieve ohmic contacts for the metal/semiconductor interfaces [67, 68].

The electrical resistivity (ρ) of the layers was calculated using Equation 3.19, where R is the electrical resistance, A is the contact area and L is the film thickness. The average electrical resistance (R) was calculated using the I–V data extracted under dark condition using the Rera Solution PV simulation system.

Figure 6.28 shows the plot of electrical resistivity (ρ) against the cathodic voltage in which the CdTe layers were grown. From the observation in Fig. 6.28, the resistivity of CdTe layers reduces after CdCl$_2$ treatment which might be due to defect and grain boundary passivation, reduction of grain boundaries due to grain growth during CdCl$_2$ treatment and increase in CdTe crystallinity amongst other advantages of CdCl$_2$ treatment as documented in the literature [27, 28].

Both the as-deposited and CdCl$_2$-treated CdTe layers grown at ~1360 mV cathodic voltage exhibit the high resistivity value. This is expected since the CdTe is stoichiometric and intrinsic in electrical conduction. The material grown at 1360 mV should, therefore, have the highest resistivity. However, cathodic voltages

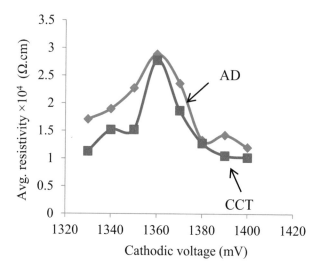

Fig. 6.28 Typical graphs of electrical resistivity against cathodic voltage for CdTe layers grown within the cathodic voltages between 1330 and 1400 mV in both as-deposited and CdCl$_2$-treated conditions

closer to the intrinsic voltage (V_i) would be favourable for the fabrication of devices due to high crystallinity, better optical and morphological properties achieved in stoichiometric or near-stoichiometric CdTe layers. It should be noted that based on the understanding as published by Dharmadasa et al. (2002) [63, 69, 70] and new results as published by independent researchers such as Reese et al. [65] and Burst et al. [64], fabricating devices with slightly Cd-rich CdTe is favourable due to increased carrier lifetime and defect reduction amongst other advantages.

Summations

In this work, CdTe layers were successfully electrodeposited using a two-electrode configuration from an acidic and aqueous solution containing cadmium chloride hydrate $CdCl_2 \cdot xH_2O$ and tellurium oxide TeO_2 as Cd and Te precursors, respectively. XRD analysis shows (111) cubic as the preferred orientation of all the CdTe explored in this work, while the highest diffraction intensity was observed at 1360 mV under both as-deposited and $CdCl_2$-treated conditions. After $CdCl_2$ treatment, an improvement in the absorption edge slope was observed with the largest slope signifying highest CdTe quality [28, 71]. This was observed at 1360 mV. Morphologically, better glass/FTO substrate coverage was observed for Cd-rich CdTe layer grown at 1400 mV, while Te-rich layer grown at 1330 mV shows high pinhole density. The best underlying substrate coverage, grain growth and size were observed at 1360 mV. PEC measurements show the ability to electroplate n-, i- and p-type CdTe layers using precursors explored in this work.

6.6.3   Effect of I-Doping on the Material Properties of CdTe

The motivation for the exploration of I-doping of CdTe is based on the new model as put forward by Dharmadasa and co-workers in 2002 [69, 72]. This understanding is associated with Te richness in CdTe and defect levels close to the conduction band, while Cd richness in CdTe is associated with defect levels close to the valence band from which higher barrier height can be achieved. Although Cd-rich CdTe can be achieved intrinsically, this work focuses on the optimisation of extrinsic iodine doping of CdTe since it is the best atomic replacement for Te atoms with minimal deformation to the CdTe lattice.

6.6.3.1   Growth of I-Doped CdTe Layers

CdTe:I thin films were electrodeposited cathodically on glass/FTO substrates by a potentiostatic technique in which the anode was a high-purity graphite rod (see Sect. 3.3). The main difference to the CdTe bath as described in Sect. 6.3 is the incorporation of the varied concentration of cadmium iodide (CdI_2) between 0 and 200 ppm with a purity of 99.995% to different CdTe baths containing the same measured salts of Cd and Te precursors. The setup and substrate preparation remain identical to the CdTe bath as described in Sect. 6.3. All the CdTe:I layers explored were grown at 1370 mV based on prior analysis as demonstrated in Sect. 6.3.

6.6.3.2 Material Properties of CdTe:I

Structural Analysis

Figure 6.29a–c shows the X-ray diffraction (XRD) spectra of CdTe:I layer under the AD, CCT and $CdCl_2+Ga_2(SO_4)_3$ treatment (GCT) conditions grown from electrolytic baths containing different I-concentrations, respectively. Figure 6.29d shows the CdTe (111) peak intensity at different post-growth treatments against I-doping concentration in the CdTe electrolytic bath. The GCT was performed by adding few drops of aqueous solution containing 0.1 M $CdCl_2$ and 0.05 M of $Ga_2(SO_4)_3$ in 20 mL of DI water to the surface of the semiconductor layer. The full coverage of the layers with the treatment solutions was achieved by spreading the solution using solution-damped cotton bud. The semiconductor layer was allowed to air-dry and heat treated at 400 °C for 20 min.

From observation, diffraction peaks associated with CdTe (111), (220) and (311) all in the cubic phase were observed aside from the diffractions associated with the glass/FTO underlying substrates. It should be noted that the extracted XRD data from this CdTe:I work matches the Joint Committee on Powder Diffraction Standards (JCPDS) reference file number 01-075-2086 on cubic CdTe layers. From the observations in Fig. 6.29a–c, no diffraction associated with elemental Iodine or Iodine related compounds were observed in all of the explored XRD layers, which might be due to the low concentrations investigated and the sensitivity of the XRD technique utilised in this work. Under all treatment conditions, the CdTe (111) diffraction shows the highest diffraction intensity which is synonymous with the preferred orientation of CdTe:I layer under the growth conditions of this study. Comparative improvement in the CdTe:I layers was observed after CCT or GCT as shown by higher (111)C CdTe XRD intensity as compared to the AD CdTe:I layers. This observation can be attributed to Cd/Te stoichiometric improvement by sublimation of excess element or the formation of CdTe from excess elements [28].

As shown in Fig. 6.29a–c, under AD, CCT and GCT conditions, an initial increase in the (111) cubic CdTe diffraction intensity was observed from 0 to 5 ppm I-doping which signifies an improvement in the CdTe:I crystallinity, while a gradual decline in the diffraction intensity of CdTe:I was observed at I-doping concentration at 10 ppm and above as clearly shown by Fig. 6.29d. The observed reduction in X-ray diffraction intensity of the (111)C peak at 10 ppm I-doping and above might be due to the formation of CdI_2 complexes such as CdI^+, CdI_2, CdI_3^- and CdI_4^{2-} in aqueous solution [73], which were not effective in nonaqueous electrolytic deposition of CdTe incorporating iodine [46]. The complexes CdI_2 formed hinder the deposition of Cd^{2+} ions on the surface of the working electrode and also reduce the available Cd^{2+} with increasing CdI_2 concentration as suggested by Paterson et al. [73, 74].

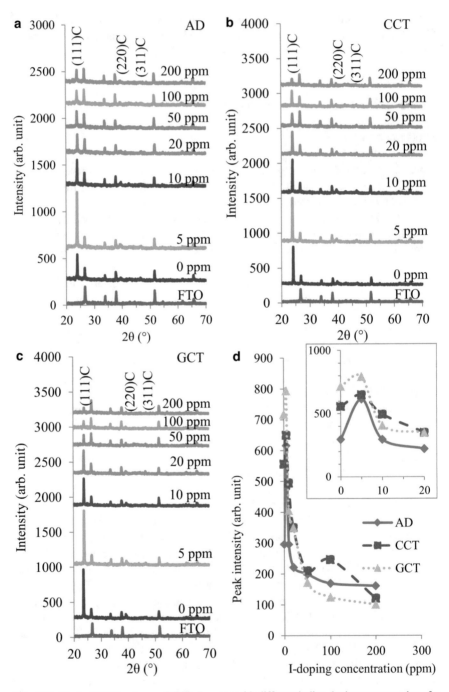

Fig. 6.29 Typical XRD patterns of CdTe:I grown with different iodine doping concentrations for (**a**) AD, (**b**) CCT and (**c**) GCT; (**d**) is the graph of (111) cubic CdTe peak intensity against I-doping concentration

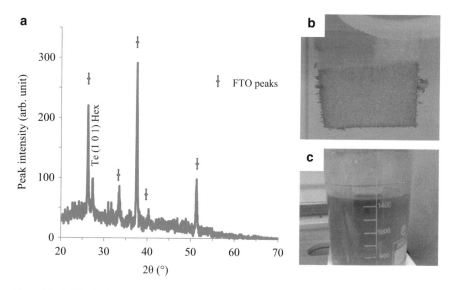

Fig. 6.30 (a) Typical XRD pattern of Te grown at 1000 ppm I-doping of CdTe, (b) deposited Te layer and (c) unstable CdTe bath with 1000 ppm I-doping

The explored I-doping concentration range was limited to 200 ppm due to further reduction in the (111) cubic CdTe intensity (without recrystallisation) and total elimination at ~1000 ppm with the appearance of the emergence of the hexagonal (101) Te peak as shown in Fig. 6.30a, while the deposited layers are characterised with low adhesion as shown in Fig. 6.30b. Figure 6.30c depicts the colouration of the electrolyte due to the instability of the electrolytic bath. This observation signifies the nondeposition of Cd on the glass/FTO substrate at high I-doping concentration.

Table 6.9 shows the (111) cubic CdTe X-ray diffraction analysis for the AD, CCT and GCT-CdTe:I layers. The crystallite sizes were calculated using Scherrer's equation as shown in Equation 3.8. For the as-deposited CdTe:I layer as shown in Table 6.9, a gradual reduction in the d-spacing was observed with increase in I-doping concentrations, while samples with a concentration above 100 ppm exhibit significant alteration in the d-spacing. This observation iterates the presence of tensile stress in the crystal plane due to the inclusion of I. Under AD conditions, a constant crystallite size of 65.38 nm was observed for CdTe:I-doped layers with I-doping concentration ranging between 0 and 20 ppm. Above this range, the crystallite size was reduced to ~52.26 nm. Similar trends of reduction in the crystallite sizes were observed under the CCT above 5 ppm I-doping and GCT above 10 ppm I-doping concentration. Alteration in crystallite stress distribution, compositional configuration, oxidation and grain growth amongst other factors may drastically affect the crystallite parameters as observed in Table 6.9. It should be noted that the stagnated crystallite sizes at 65.38 and 52.26 nm might be due to the limitation of the use of the Scherrer's equation for materials with larger grains as well as the XRD machine [59, 75].

Table 6.9 The summary of the X-ray diffraction analysis for (111) cubic CdTe diffraction

Doping (ppm)	2θ (°)	D-spacing (Å)	FWHM (°)	Crystallite size (nm)
AD				
0	23.92	3.71	0.1298	65.4
5	23.91	3.71	0.1298	65.4
10	23.97	3.70	0.1298	65.4
20	23.98	3.70	0.1298	65.4
50	23.97	3.70	0.1624	52.3
100	24.06	3.69	0.1624	52.3
200	24.03	3.69	0.1624	52.3
CCT				
0	23.94	3.71	0.1298	65.4
5	24.11	3.68	0.1298	65.4
10	23.89	3.72	0.1623	52.3
20	23.95	3.71	0.1623	52.3
50	23.94	3.71	0.1624	52.3
100	23.95	3.71	0.1623	52.3
200	23.95	3.71	0.2597	32.7
GCT				
0	23.9	3.72	0.1298	65.4
5	23.99	3.70	0.1298	65.4
10	24.07	3.69	0.1298	65.4
20	23.6	3.76	0.1623	52.3
50	23.92	3.71	0.1623	52.3
100	23.98	3.70	0.1623	52.3
200	23.92	3.71	0.1623	52.3

Optical Properties Analysis

The optical absorbance measurement of the CdTe:I under AD, CCT and GCT conditions was performed at room temperature. The bandgaps of the CdTe:I layers were determined using the graphical plot of A^2 against photon energy ($h\nu$). Figure 6.31a, b and c shows A^2 against $h\nu$ plot of the AD, CCT and GCT-CdTe:I at different I-doping concentrations, respectively. Figure 6.31d shows a graph of absorption edge slope against I-doping concentrations of CdTe baths, and Table 6.10 shows the optical absorption properties of the CdTe:I layers under all the conditions explored in this work. From the observation, the bandgaps for the explored I-doping concentrations from 0 to 200 ppm fall within the range of 1.52 ± 0.05 eV for the AD CdTe:I layer with an increase in the I-doping concentration resulting in increase in the bandgap. Both the CCT and GCT treated CdTe layers fall within the bandgap range of 1.48 ± 0.01 and 1.47 ± 0.02 eV, respectively. The improvement towards the acceptable CdTe bandgap range of 1.44–1.50 eV [76] and the narrowing of the bandgap range are due to the enhancement of material properties such as grain growth, crystallinity and sublimation of excess element [28, 77, 78] amongst others.

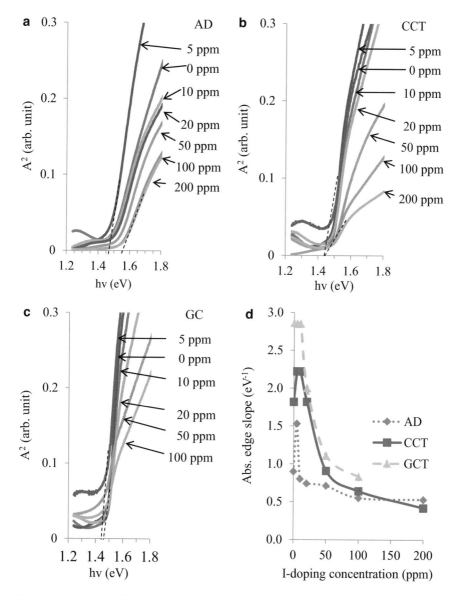

Fig. 6.31 Optical absorption of CdTe:I with different doping concentrations of iodine under (**a**) AD, (**b**) CCT and (**c**) GCT conditions, while (**d**) is the absorption edge slope of CdTe:I under AD, CCT and GCT conditions against I-doping concentration

Furthermore, the optical absorption edge slope which is a quantitative measure of defect and impurity energy levels [28, 71] was explored as a function of I-doping concentration as shown in Fig. 6.31d. Under all the post-growth treatment conditions investigated in this work, the highest absorption edge slope was observed at 5 ppm I-doping concentration. This comparatively signifies that more incident photons can

Table 6.10 The optical bandgaps and slopes of absorption edge of CdTe layers grown from different CdTe baths with different I-doping concentrations having undergone different post-growth treatments

I-doping (ppm)	Bandgap (eV)			Abs. edge slope (eV^{-1})		
	AD	CCT	GCT	AD	CCT	GCT
0	1.47	1.47	1.46	0.90	1.82	2.86
5	1.47	1.48	1.49	1.53	2.22	2.85
10	1.49	1.49	1.49	0.80	2.22	2.85
20	1.50	1.49	1.48	0.74	1.82	2.00
50	1.53	1.49	1.45	0.71	0.91	1.11
100	1.55	1.49	1.45	0.55	0.64	0.83
200	1.56	1.49		0.53	0.42	

be absorbed in a few nanometer of CdTe:I thickness and also the increased possibility of achieving higher solar-to-electricity conversion efficiency when incorporated in solar cell structures.

Morphological Property Analysis

Figure 6.32 shows the SEM micrographs of CdTe:I incorporating (a) 5 ppm I-doping under CCT and (b) under GCT, while (c) incorporating 100 ppm I-doping under CCT and (d) under GCT. The as-deposited CdTe:I and 0 ppm I-doped CdTe layers were not incorporated due to its triviality.

In accord with the effect of CCT and GCT as documented in the literature, grain growths were observable as compared to the as-deposited CdTe:I layers. Comparatively, a reduction in the grain size was observed above 5 ppm I-doping of the CdTe bath which may be due to competing phases of CdI$_2$ complexes in aqueous solution as suggested in the literature [73, 74]. This observation is corroborative with the structural and optical analytical observations.

Photoelectrochemical (PEC) Cell Measurement

Figure 6.33 shows the graph of PEC cell measurement against varying I-doping concentration from 0 to 200 ppm in the CdTe:I bath. From the observation, all the as-deposited CdTe:I layers at all the explored I-doping concentrations show n-type conductivity as it is well known that iodine along with other halogens is a known donor to CdTe due to the introduction of excess electrons into the conduction band [79]. For both the CCT- and GCT-treated CdTe:I layers, the initial n-type conductivity was retained from the as-deposited CdTe:I for layers between 0 and 10 ppm I-doping concentration, while the transition to p-type conductivity was observed above 10 ppm I-doping concentration. It should be noted that the conduction type of a semiconductor material depends on the domination of factors such as elemental composition, doping alteration due to annealing parameters and defect distribution amongst others [23, 80].

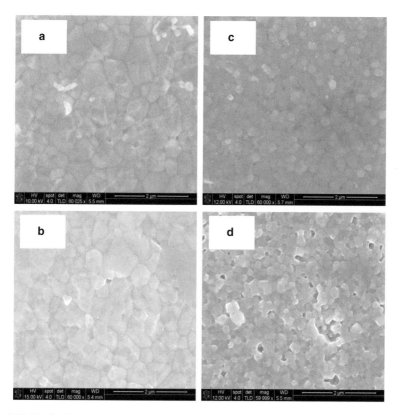

Fig. 6.32 Typical SEM micrographs of CdTe:I (**a**) incorporating 5 ppm I-doping under CCT and (**b**) under GCT, while (**c**) incorporating 100 ppm I-doping under CCT and (**d**) under GCT

DC Properties Analysis

The formation of ohmic contacts with the CdTe:I was dependent on the conductivity type of the layers as discussed in the PEC section. Au was evaporated on p-type layers, while In was evaporated on n-type layers. Figure 6.34 shows the graph of electrical conductivity against I-doping concentration in the CdTe bath. Relatively, a continuous reduction in the conductivity of the deposited layers with increasing I-doping concentration was observed with the exception of the CdTe:I layer grown at 5 ppm I-doping. This observation signifies a possible solubility of CdTe:I at 5 ppm in aqueous solution. Other factors to be considered for this trend include the comparatively higher mobility and the conductivity of n-type semiconductors relative to their p-type counterpart as documented in the literature [81] and the possible effect of formation of complexes in aqueous solution [73, 74].

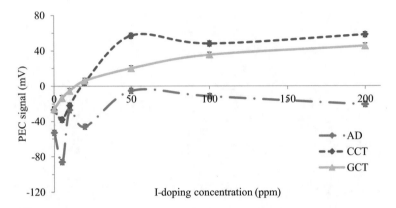

Fig. 6.33 Photoelectrochemical cell measurement for AD, CCT and GCT-CdTe:I layers

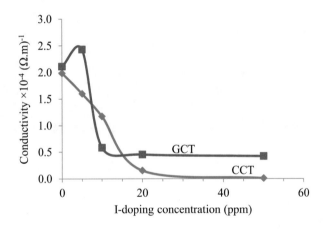

Fig. 6.34 Graph of electrical conductivity against I-doping concentration in CdTe bath

Summations

In this work, the effect of iodine doping in a CdTe electrolytic bath on the CdTe layer as it affects its structural, optical, morphological and compositional properties was explored. An optimal I-doping concentration of 5 ppm in the electrolytic bath was observed under all material characterisations explored in this work due to the possible solubility limit achieved at 5 ppm I-doping and deterioration in material property afterwards which is contrary to the improvement of material and electronic properties as documented in the literature for electrodeposited CdTe:I in nonaqueous electrolytes [46].

6.6.4   Effect of Ga-Doping on the Material Properties of CdTe

The motivation for the in situ incorporation of Ga in CdTe is based on the possibility of the reduction of Te precipitation as demonstrated by the literature after post-growth treatment [24, 42]. This has also been demonstrated by the presence of Cd [42] and Cl [43] in the literature to reduce the Te precipitation [21]. Therefore, the effect of in situ Ga-doping with different Ga-doping concentrations of CdTe:Ga has been explored.

6.6.4.1   Growth of Ga-Doped CdTe Layers

CdTe:Ga thin films were electrodeposited cathodically on glass/FTO substrates by a potentiostatic technique in which the anode was a high-purity graphite rod (see Sect. 3.3). The main difference to the CdTe bath as described in Sect. 6.3 is the incorporation of varied concentrations from 0 to 200 ppm of gallium sulphate ($Ga_2(SO_4)_3$) with a purity of 99.99% to different CdTe baths containing the same measured salts of Cd and Te precursors. The setup and substrate preparation remains identical to the CdTe bath as described in Sect. 6.3. All the CdTe:Ga layers explored were grown at 1370 mV based on prior analysis as demonstrated in Sect. 6.3.

6.6.4.2   Material Properties of CdTe:Ga

Structural Analysis

The structural analysis was performed to determine the optimal Ga-doping concentration in the CdTe bath. This can be determined by observing the highest peak intensity, crystallinity and crystallite size. Figure 6.35a–c shows the X-ray diffraction (XRD) measured between $2\theta = 20°$ and $2\theta = 70°$ for CdTe layers doped with Ga between 0 and 200 ppm under the AD, CCT and GCT conditions. Figure 6.35d shows the (111) cubic CdTe XRD intensity of CdTe grown from electrolyte containing different concentrations of Ga-doping ranging from 0 to 200 ppm. Aside from the XRD diffraction associated with glass/FTO at ~26.64°, ~33.77°, ~37.85°, ~51.64° and ~65.62° 2θ angle, CdTe peaks associated with (111) cubic, (220) cubic and (311) cubic were observed at $2\theta \approx 23.95°$, $2\theta \approx 38.60°$ and $2\theta \approx 45.80°$, respectively. From the observation, the most intense diffraction, which signifies the preferred orientation of CdTe growth, was observed at $2\theta \approx 23.95°$ for all the post-growth treatment conditions explored in this work. Similarly, a gradual increase in the diffraction intensity was observed with increasing Ga-doping from 0 to 20 ppm as shown in Fig. 6.35a–d. This observation can be associated with an increase in crystallinity of CdTe along the (111)C plane due to the reduction and dissolution of Te precipitation [24] and the incorporation of Ga as a substitutional dopant. An increase in the doping of Ga above 20 ppm results in a

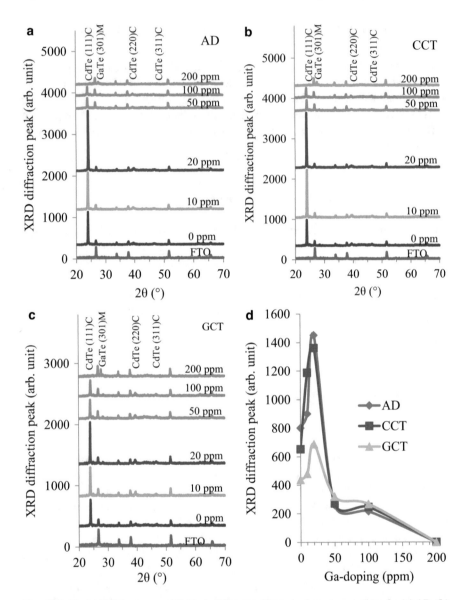

Fig. 6.35 Typical XRD patterns of CdTe at different gallium-doping concentrations for (**a**) AD, (**b**) CCT and (**c**) GCT-CdTe layers. (**d**) is a typical plot of CdTe:Ga (111) cubic peak intensity against Ga-doping of CdTe baths

reduction in diffraction intensity of the CdTe (111)C and a total collapse of the peak at 200 ppm Ga-doping and above.

Subsequently, an emergence of a GaTe diffraction peak was observable above 20 ppm Ga-doping under all the conditions explored in this work. Furthermore, an increase in the GaTe peak intensity at $2\theta \approx 27.23°$ was noticeable with increasing Ga

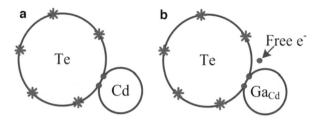

Fig. 6.36 Covalent bond formation between (**a**) Cd and Te atoms and (**b**) Ga occupying Cd sites and bonding with Te as a result of gallium incorporation

concentration. It should be noted that the reduction of CdTe (111)C diffraction intensity above 20 ppm Ga-doping in the CdTe bath can be attributed to competing for crystalline phases of CdTe and GaTe. Furthermore, this observation emphasises the replacement of Cd atoms with Ga atoms in the crystal lattice to form GaTe [42]. Only two valence electrons out of three available for bonding of Ga atoms were utilised in the formation of bonds with neighbouring Te atoms, while the excess electron is donated to the lattice to aid the *n*-type conduction (see Fig. 6.36).

The extracted XRD data from this work matches the Joint Committee on Powder Diffraction Standards (JCPDS) reference file number 01-075-2086-cubic for CdTe and 01-075-2220 monoclinic for GaTe. Table 6.11 shows the XRD analysis of CdTe layers grown at different gallium concentrations. The crystallite sizes (*D*) of each of the CdTe layers at different Ga-doping were calculated using Scherrer's formula as shown in Equation 3.8.

Uniformity in both the FWHM and crystallite size was observed for the entire as-deposited CdTe layers as shown in Table 6.11. After CCT treatment of the CdTe layers, reduction in FWHM to 0.129° and resulting increase in crystallite sizes to 65.3 nm were observed for CdTe layers doped with Ga within the range of 0 and 20 ppm. Subsequently, reductions in the crystallite size were observed for CdTe layers doped with Ga at 50 ppm and above. This observation is due to the reduction of crystallinity due to the formation of two phases: CdTe and GaTe. The replacement of Cd atoms which have an atomic radius of 1.61 Å with Ga atoms which have a comparatively lower atomic radius of 1.36 Å may lead to contraction of the crystal lattice [57, 77]. The same phenomenon was observed with the GCT-treated CdTe:Ga layers.

Furthermore, it is appropriate to look at the stability of the two compounds CdTe and GaTe at this stage. The stability of inorganic compounds is given by their heat of formation or the enthalpy value (ΔH) given by Equations 6.4 and 6.5.

$$\text{(reactant)} \quad Cd + Te \rightarrow CdTe \ \text{(product)} + \Delta H \qquad \text{(Equation 6.4)}$$

$$\text{(reactant)} \quad Ga + Te \rightarrow GaTe \ \text{(product)} + \Delta H \qquad \text{(Equation 6.5)}$$

The product with the most significant negative value of ΔH is more stable and has a high tendency to proceed [82]. As depicted in Fig. 6.37, the ΔH values of CdTe and GaTe are -50.6 and -62.6 kJmol^{-1} [83, 84], respectively, as reported in the

Table 6.11 Summary of XRD analysis of CdTe layers grown with different Ga-doping concentrations

Doping (ppm)	2θ (°)	d-spacing (Å)	FWHM (°)	Crystallite size (nm)
AD				
0	23.95	3.712	0.162	52.3
10	23.95	3.712	0.162	52.3
20	24.12	3.688	0.162	52.3
50	23.94	3.714	0.162	52.3
100	23.95	3.712	0.162	52.3
CCT				
0	24.10	3.689	0.130	65.3
10	23.95	3.713	0.130	65.3
20	23.90	3.720	0.130	65.3
50	23.98	3.708	0.162	52.3
100	23.93	3.716	0.162	52.3
GCT				
0	23.94	3.714	0.130	65.3
10	23.99	3.707	0.162	52.3
20	23.94	3.714	0.162	52.3
50	23.96	3.710	0.162	52.3
100	24.08	3.693	0.195	43.6

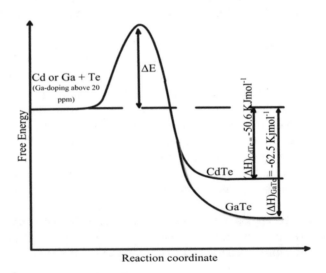

Fig. 6.37 Free energy graph depicting the enthalpy ΔH_f at 298 K for both CdTe and GaTe compounds

literature. These values indicate that GaTe is more stable, and therefore, as the Ga concentration increases beyond 20 ppm, GaTe compound formation takes place within the layer. The presence of these two competing phases allows the deterioration of crystallinity and the complete collapse of the (111)C peak originating from CdTe.

Optical Properties Analysis

Figure 6.38a–c shows the optical absorption of CdTe:Ga layers under the AD, CCT and GCT conditions, respectively, while Fig. 6.38d shows the graph of absorption edge slope against Ga-doping concentration in CdTe bath. It should be noted that due to the removal of the effect of glass/FTO as explained in Sect. 3.4.6, only the absorption of the CdTe:Ga is captured. As observed in Fig. 6.38a–c, the optical bandgaps of the entire deposited CdTe layer at varying Ga-doping concentration were within the range of 1.48 ± 0.02 eV. This observation shows the dominance of CdTe even with the incorporation of Ga and formation of GaTe as observed in the structural analysis section. Alteration in the absorption edges of CdTe:Ga layers with varying Ga-doping concentrations was also observed as shown in Fig. 6.38d. The steepest absorption edge slope which signifies the best quality of semiconductor layer [28] was observed at a Ga-doping of 20, 10 and 0 ppm under the AD, CCT and GCT conditions, respectively (Fig. 6.38d). As reported in the literature, increase in steepness of the semiconductor materials absorption edge signifies lesser defects and impurity energy levels in the thin film [85]. Furthermore, a sharp absorption edge also allows more photons to be absorbed even at low CdTe thickness of a few microns [28]. This optical result, therefore, shows the possibility of having better solar cell efficiency using 20 ppm gallium-doped CdTe:Ga.

This observation indicates an initial improvement in the material quality for CdTe:Ga layers with Ga-doping of 0–20 ppm under AD and CCT conditions as shown in Fig. 6.38d and also suggests CdTe:Ga material quality saturation at 20 ppm Ga-doping under the conditions explored in this set of experiments. An increase in Ga-doping above 20 ppm for CdTe:Ga layers under AD and CCT shows the detrimental effect to the CdTe:Ga material quality as the reduction in the absorption edge slope suggests for Ga-doping of 50 ppm and above. This observation is in accord with the structural analysis as described earlier.

Furthermore, a dissimilar trend of absorption edge slope for the GCT-treated CdTe:Ga layers was observed with undoped layers doped with 0 ppm giving the highest absorption edge slope. Although, it is well known that the treatment of CdTe in the presence of Ga melt improves the material properties of CdTe by dissolving Te precipitates [24, 42], the incorporation of superfluous Ga into the CdTe lattice results in detrimental effects such as the formation of competing phases of CdTe and GaTe as observed in the structural analysis section. It should be noted that the effects of persistent photoconductivity and photo-induced persistent absorption associated with Ga-doped CdTe at 77 K [86] were not observed due to the room temperature at which the experiments were performed.

Morphological Analysis

Figure 6.39a–l shows the SEM micrographs of CdTe layers grown from electrolytes containing 0, 20, 50 and 100 ppm Ga-doping under the AD, CCT and GCT conditions. From observation, the initial incorporation of Ga-doping from 0 to

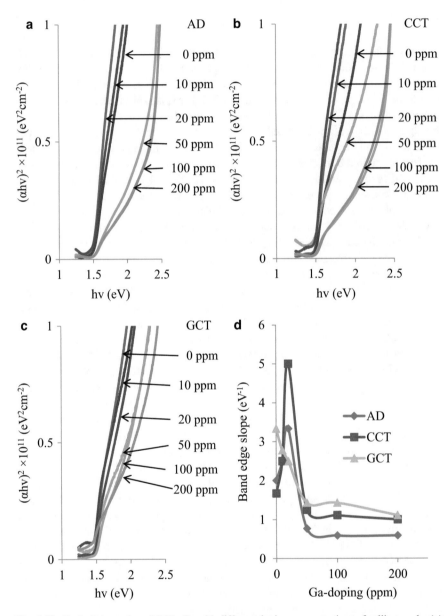

Fig. 6.38 Optical absorption of CdTe:Ga with different doping concentrations of gallium under (**a**) AD, (**b**) CCT and (**c**) GCT conditions, while (**d**) is the absorption edge slope of CdTe:Ga under AD, CCT and GCT conditions against Ga-doping concentration

20 ppm under AD condition shows agglomeration of grains and full coverage of the underlying glass/FTO substrate. At 50 ppm Ga-doping, a reduction in grain and agglomeration sizes was observed, while no clear grain boundaries were observed at 100 ppm and above. The deterioration in grain and agglomeration

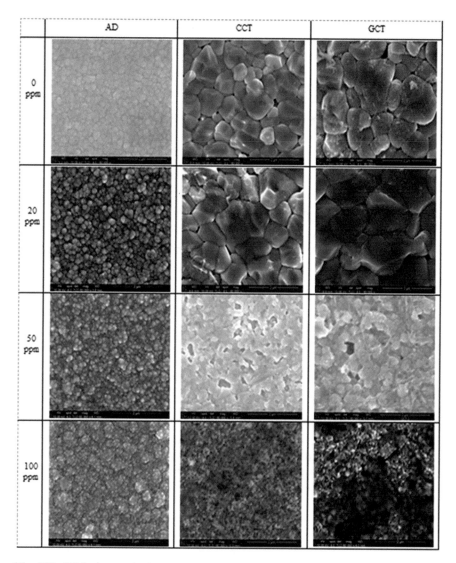

Fig. 6.39 SEM micrographs for CdTe:Ga layers grown with Ga-doping concentrations between 0 and 100 ppm which have undergone AD, CCT and GCT treatments

sizes can be attributed to competing crystalline phases of CdTe and GaTe. As expected, improvement after CCT and GCT post-growth treatment was observed with larger grain size at 0–20 ppm. The treatment of CdTe doped with Ga higher than 20 ppm shows detrimental effect such as comparatively smaller grains and high pinhole density. It should be noted that the pinholes observed in the 0–20 ppm Ga CdTe layers might be due to optimisation requirement for both heat treatment temperature and duration, Ostwald ripening, the thickness of the CdTe layer utilised and the utilisation of glass/FTO substrate. It is well known that CdTe grows better on smooth semiconductor (with CdS being a preferred partner) surfaces due to its columnar growth configuration. Based on this observation, it could be said that the superfluous incorporation of Ga through in situ Ga-doping of CdTe or GCT is detrimental to CdTe layer property.

This observation is in accord with the structural and optical analyses discussed in the structural and optical analysis sections. It should be noted that the crystallite size as calculated using XRD does not correspond to the grain size as seen on the SEM micrograph but the grains are formed from the agglomeration of many crystallites.

Photoelectrochemical (PEC) Cell Measurement

Figure 6.40 shows the photoelectrochemical cell measurement against Ga-doping concentration in CdTe bath. The CdTe layers utilised for these experiments were grown at 1370 mV. For the as-deposited CdTe:Ga layers, a retention of the n-type conductivity was maintained from 0 to 20 ppm, while a transition from n- to p-type was observed at 50 ppm and above. The change from n- to p-type conductivity might be due to the excess replacement of Cd with Ga, whereby the doping effect is converted into an alloying effect with the complete alteration of the crystallite lattice and compound.

The excess incorporation of Ga at 50 ppm and above might create an acceptor-like defect which might be more dominant than the n-doping characteristics of Ga in CdTe. Furthermore, the formation of GaTe which is typically a p-type semiconductor material [87, 88] as observed in structural analysis section might also be the possible cause. After both CCT and GCT, the conductivity type of CdTe:Ga at all doping concentrations explored shows an n-type electrical conductivity due to the changes occurring in the CdTe:Ga layers. This might be through the sublimation of excess element and the removal of acceptor-like defects during heat treatment.

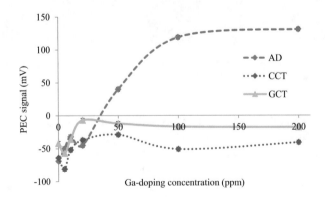

Fig. 6.40 Photoelectrochemical cell measurements for AD-, CCT- and GCT-treated Ga-doped CdTe thin films

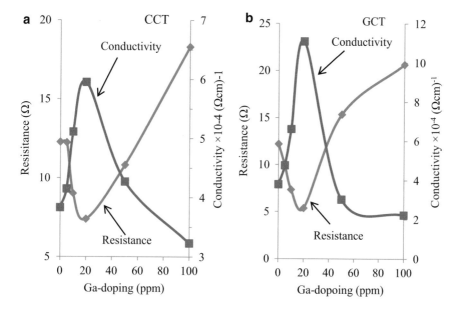

Fig. 6.41 Electrical resistance and conductivity plotted against Ga-doping concentration in the CdTe bath after (**a**) CCT and (**b**) GCT treatments

This conductivity-type change after heat treatment indicates that the n-doping is dominant rather than any p-like native defects.

DC Conductivity Properties Analysis

Figure 6.41a, b shows the graph of both the resistance and conductivity against Ga-doping concentration of CdTe after CCT and GCT post-growth treatment conditions. For this experiment, ~1.5-μm-thick CdTe:Ga layers grown from CdTe baths with varied Ga concentrations from 0 to 100 ppm were utilised. The glass/FTO/CdTe:Ga layers were cut into two, and each half was either treated with CCT or GCT. Two-millimeter diameter indium ohmic contacts with a thickness of ~100 nm were evaporated on the treated CdTe:Ga layers. It should be noted that only the CCT and the GCT post-growth treated CdTe:Ga was explored due to its significance in the determination of device properties.

The electrical resistance (R) of the structure was calculated from the ohmic I–V data obtained under dark condition using a Rera Solution PV simulation system using the glass/FTO/CdTe:Ga/In. The resistivity values were obtained using Equation 3.19.

The comparable trend in the CCT- and the GCT-treated CdTe:Ga layers was observed as shown in Fig. 6.41. The incorporation of Ga-doping from 0 ppm shows a gradual increment in conductivity which can be attributed to the excess electrons donated from the replacement of Cd atoms with Ga atoms in the crystal lattice

[42]. A possible solubility limit of CdTe:Ga was observed at ~20 ppm Ga-doping due to the gradual reduction in the conductivity [54] beyond 20 ppm doping in all conditions observed. The domination of p-type conduction as discussed in the PEC analysis is due to the superfluous incorporation of Ga-incorporation above 20 ppm. The change in the conduction of CdTe from n- to p-type at high Ga-doping might also be a determining factor in the observed conductivity trend as shown in Fig. 6.41, as it is well known that the mobility and the conductivity of p-type semiconductors are lower than that of their n-type counterpart [81].

Summations

The work presented in this section shows the successful exploration of the effect of in situ Ga-doping concentration in the CdTe electrolytic bath and the inclusion of Ga in the regular $CdCl_2$ post-growth treatment on electrodeposited CdTe layers. The XRD results show an increase in the preferred (111) cubic CdTe peak from 0 to 20 ppm Ga-doping for all the post-growth treatments explored. A gradual reduction in the (111) cubic CdTe peak was observed with an emergence of the (301) monoclinic GaTe peak at 50 ppm Ga-doping and above, coupled with a reduction in crystallite size. The optical analysis shows improvement in the absorption edge with in situ Ga-doping of CdTe from 0 to 20 ppm. Morphologically, a comparative reduction in the grain size was observed at high in situ Ga-doping at 50 ppm and above. PEC cell measurement shows a transition of conductivity type from n-type to p-type at high Ga-doping of 50 ppm and above for the as-deposited CdTe layers.

6.7   Conclusions

The work presented in this chapter demonstrates the electrodeposition of CdTe using two-electrode configuration from an electrolytic aqueous bath containing cadmium nitrate tetrahydrate $Cd(NO_3)_2 \cdot 4H_2O$ and tellurium oxide (TeO_2) as cadmium and tellurium precursors, respectively. Based on the material characterisation techniques explored, 1370 mV was identified as the best cathodic potential in which stoichiometric CdTe is achieved. The XRD results show the presence of cubic CdTe both in the AD and post-growth treated CdTe layer with a preferred orientation along the (111) plane. Both the XRD and SEM results in this work indicate grain growth after $CdCl_2$ treatment with an increase in crystallite sizes and the formation of large grains within the range of 300–2000 nm. The optical absorption results showed a bandgap range of 1.45 and 1.50 eV after $CdCl_2$ treatment.

The effects of different dopants (F-, I-, Ga- and Cl-based precursor) in the CdTe electrolytic bath were also explored.

References

1. M.P.R. Panicker, M. Knaster, F.A. Kroger, Cathodic deposition of CdTe from aqueous electrolytes. J. Electrochem. Soc. **125**, 566 (1978). https://doi.org/10.1149/1.2131499
2. T.M. Razykov, C.S. Ferekides, D. Morel, E. Stefanakos, H.S. Ullal, H.M. Upadhyaya, Solar photovoltaic electricity: current status and future prospects. Sol Energy **85**, 1580–1608 (2011). https://doi.org/10.1016/j.solener.2010.12.002
3. K. Zanio, *Semiconductors and Semimetals* (Academic, New York, 1978). http://shu.summon. serialssolutions.com/2.0.0/link/0/eLvHCXMwdV3JCsIwEB1cEAQPrrgV-gNKmyZNPYvFu9 4l6bQ3K1j_HydDXXA5Zg7DJJB5me0FIBLrYPXhE8LEUZ8lRggjsgBlgBuptSqw0Chzrsy8 0Rg848ZXCuObQZ_iCKmCyN3HJjQJON2LqOaiYzdM7pnQiml0hCaUCiNVM-481sn7lwY MKGkfWm7IYACNvBxCh9sws2o
4. J.J. Loferski, Theoretical considerations governing the choice of the optimum semiconductor for photovoltaic solar energy conversion. J. Appl. Phys. **27**, 777–784 (1956). https://doi.org/10. 1063/1.1722483
5. C.S. Ferekides, U. Balasubramanian, R. Mamazza, V. Viswanathan, H. Zhao, D.L. Morel, CdTe thin film solar cells: device and technology issues. Sol Energy **77**, 823–830 (2004). https://doi.org/10.1016/j.solener.2004.05.023
6. K.L. Chopra, P.D. Paulson, V. Dutta, Thin-film solar cells: an overview. Prog. Photovolt. Res. Appl. **12**, 69–92 (2004). https://doi.org/10.1002/pip.541
7. I.M. Dharmadasa, *Advances in Thin-Film Solar Cells* (Pan Stanford, Singapore, 2013)
8. O.K. Echendu, F. Fauzi, A.R. Weerasinghe, I.M. Dharmadasa, High short-circuit current density CdTe solar cells using all-electrodeposited semiconductors. Thin Solid Films **556**, 529–534 (2014). https://doi.org/10.1016/j.tsf.2014.01.071
9. P.V. Braun, P. Osenar, M. Twardowski, G.N. Tew, S.I. Stupp, Macroscopic nanotemplating of semiconductor films with hydrogen-bonded lyotropic liquid crystals. Adv. Funct. Mater. **15**, 1745–1750 (2005). https://doi.org/10.1002/adfm.200500083
10. T. Nishio, M. Takahashi, S. Wada, T. Miyauchi, K. Wakita, H. Goto, S. Sato, O. Sakurada, Preparation and characterization of electrodeposited in-doped CdTe semiconductor films. Electr Eng Japan **164**, 12–18 (2008). https://doi.org/10.1002/eej.20673
11. H.I. Salim, V. Patel, a. Abbas, J.M. Walls, I.M. Dharmadasa, Electrodeposition of CdTe thin films using nitrate precursor for applications in solar cells. J. Mater. Sci. Mater. Electron. **26**, 3119–3128 (2015). https://doi.org/10.1007/s10854-015-2805-x
12. N.A. Abdul-Manaf, H.I. Salim, M.L. Madugu, O.I. Olusola, I.M. Dharmadasa, Electro-plating and characterisation of CdTe thin films using CdCl2 as the cadmium source. Energies **8**, 10883–10903 (2015). https://doi.org/10.3390/en81010883
13. S. Bonilla, E.A. Dalchiele, Electrochemical deposition and characterization of CdTe polycrystalline thin films. Thin Solid Films **204**, 397–403 (1991). https://doi.org/10.1016/0040-6090 (91)90078-C
14. C.G. Morris, *Academic Press Dictionary of Science and Technology* (Academic, San Diego, 1991)
15. V. Petr, *Electrochemical Series, CRC Handbook of Chemistry and Physics*, 87th edn. (CRC Press, Boca Raton, 2005), pp. 1–10. https://doi.org/10.1136/oem.53.7.504
16. F.A. Kröger, Cathodic deposition and characterization of metallic or semiconducting binary alloys or compounds. J. Electrochem. Soc. **125**, 2028 (1978). https://doi.org/10.1149/1. 2131357
17. J.H. Chen, C.C. Wan, Dependence of the composition of CdTe semiconductor on conditions of electrodeposition. J. Electroanal. Chem. **365**, 87–95 (1994). http://www.scopus.com/inward/ record.url?eid=2-s2.0-0040171133&partnerID=40& md5=7f6ef88803b168878b60e48117cc31a6
18. T.L. Chu, S.S. Chu, F. Firszt, H.a. Naseem, R. Stawski, Deposition and characterization of p-type cadmium telluride films. J. Appl. Phys. **58**, 1349–1355 (1985). https://doi.org/10.1063/1. 336106

19. Y. Jung, G. Yang, S. Chun, D. Kim, J. Kim, Post-growth CdCl2 treatment on CdTe thin films grown on graphene layers using a close-spaced sublimation method. Opt. Express **22**, A986–A991 (2014). http://www.scopus.com/inward/record.url?eid=2-s2.0-84899786471&partnerID=40&md5=2f630c658c679ff9e4e38e7b57ea7243

20. C. Frausto-Reyes, J.R. Molina-Contreras, C. Medina-Gutiérrez, S. Calixto, CdTe surface roughness by Raman spectroscopy using the 830 nm wavelength. Spectrochim Acta A Mol Biomol Spectrosc **65**, 51–55 (2006). https://doi.org/10.1016/j.saa.2005.07.082

21. I.M. Dharmadasa, O.K. Echendu, F. Fauzi, N.A. Abdul-Manaf, O.I. Olusola, H.I. Salim, M.L. Madugu, A.A. Ojo, Improvement of composition of CdTe thin films during heat treatment in the presence of CdCl2. J. Mater. Sci. Mater. Electron. **28**, 2343–2352 (2017). https://doi.org/10.1007/s10854-016-5802-9

22. M. Ichimura, F. Goto, E. Arai, Structural and optical characterization of CdS films grown by photochemical deposition. J. Appl. Phys. **85**, 7411 (1999). https://doi.org/10.1063/1.369371

23. B.M. Basol, Processing high efficiency CdTe solar cells. Int J Sol Energy **12**, 25–35 (1992). https://doi.org/10.1080/01425919208909748

24. P. Fernández, Defect structure and luminescence properties of CdTe based compounds. J Optoelectron Adv Mater **5**, 369–388 (2003)

25. J. Sun, D.K. Zhong, D.R. Gamelin, Composite photoanodes for photoelectrochemical solar water splitting. Energ. Environ. Sci. **3**, 1252 (2010). https://doi.org/10.1039/c0ee00030b

26. H.-G. Junginger, Electronic band structure of tellurium. Solid State Commun. **5**, 509–511 (1967). https://doi.org/10.1016/0038-1098(67)90534-0

27. I.M. Dharmadasa, Review of the CdCl2 treatment used in CdS/CdTe thin film solar cell development and new evidence towards improved understanding. Coatings **4**, 282–307 (2014). https://doi.org/10.3390/coatings4020282

28. A. Bosio, N. Romeo, S. Mazzamuto, V. Canevari, Polycrystalline CdTe thin films for photovoltaic applications. Prog. Cryst. Growth Charact. Mater. **52**, 247–279 (2006). https://doi.org/10.1016/j.pcrysgrow.2006.09.001

29. I.M. Dharmadasa, O.K. Echendu, F. Fauzi, N.A. Abdul-Manaf, H.I. Salim, T. Druffel, R. Dharmadasa, B. Lavery, Effects of CdCl2 treatment on deep levels in CdTe and their implications on thin film solar cells: a comprehensive photoluminescence study. J. Mater. Sci. Mater. Electron. **26**, 4571–4583 (2015). https://doi.org/10.1007/s10854-015-3090-4

30. M. Bugar, E. Belas, R. Grill, J. Prochazka, S. Uxa, P. Hlidek, J. Franc, R. Fesh, P. Hoschl, Inclusions elimination and resistivity restoration of CdTe:Cl Crystals by two-step annealing. IEEE Trans. Nucl. Sci. **58**, 1942–1948 (2011). https://doi.org/10.1109/TNS.2011.2159394

31. N.B. Chaure, S. Bordas, A.P. Samantilleke, S.N. Chaure, J. Haigh, I.M. Dharmadasa, Investigation of electronic quality of chemical bath deposited cadmium sulphide layers used in thin film photovoltaic solar cells. Thin Solid Films **437**, 10–17 (2003). https://doi.org/10.1016/S0040-6090(03)00671-0

32. T.L. Chu, S.S. Chu, Thin film II–VI photovoltaics. Solid State Electron. **38**, 533–549 (1995). https://doi.org/10.1016/0038-1101(94)00203-R

33. M.R. Murti, K.V. Reddy, Grain boundary effects on the carrier mobility of polysilicon. Phys Status Solidi **119**, 237–240 (1990). https://doi.org/10.1002/pssa.2211190128

34. R.V. Muniswami Naidu, A. Subrahmanyam, A. Verger, M.K. Jain, S.V.N. Bhaskara Rao, S.N. Jha, D.M. Phase, Grain boundary carrier scattering in ZnO thin films: a study by temperature-dependent charge carrier transport measurements. J Electron Mater **41**, 660–664 (2012). https://doi.org/10.1007/s11664-012-1907-y

35. E. Regalado-Pérez, M.G. Reyes-Banda, X. Mathew, Influence of oxygen concentration in the CdCl2 treatment process on the photovoltaic properties of CdTe/CdS solar cells. Thin Solid Films **582**, 134–138 (2015). https://doi.org/10.1016/j.tsf.2014.11.005

36. Z.H. Chen, C.P. Liu, H.E. Wang, Y.B. Tang, Z.T. Liu, W.J. Zhang, S.T. Lee, J.A. Zapien, I. Bello, Electronic structure at the interfaces of vertically aligned zinc oxide nanowires and sensitizing layers in photochemical solar cells. J. Phys. D Appl. Phys. **44**, 325108 (2011). https://doi.org/10.1088/0022-3727/44/32/325108

37. I.M. Dharmadasa, J. Haigh, Strengths and advantages of electrodeposition as a semiconductor growth technique for applications in macroelectronic devices. J. Electrochem. Soc. **153**, G47 (2006). https://doi.org/10.1149/1.2128120

38. A.A. Ojo, I.M. Dharmadasa, 15.3% efficient graded bandgap solar cells fabricated using electroplated CdS and CdTe thin films. Sol. Energy. **136**, 10–14 (2016). https://doi.org/10.1016/j.solener.2016.06.067

39. S. Lalitha, R. Sathyamoorthy, S. Senthilarasu, a. Subbarayan, K. Natarajan, Characterization of CdTe thin film - dependence of structural and optical properties on temperature and thickness. Sol. Energy Mater. Sol. Cells. **82**, 187–199 (2004). https://doi.org/10.1016/j.solmat.2004.01.017

40. T.L. Chu, S.S. Chu, Y. Pauleau, K. Murthy, E.D. Stokes, P.E. Russell, Cadmium telluride films on foreign substrates. J. Appl. Phys. **54**, 398 (1983). https://doi.org/10.1063/1.331717

41. K. Peters, A. Wenzel, P. Rudolph, The p-T-x projection of the system Cd-Te. Cryst Res Technol **25**, 1107–1116 (1990). https://doi.org/10.1002/crat.2170251002

42. N.V.V. Sochinskii, V.N.N. Babentsov, N.I.I. Tarbaev, M.D. Serrano, E. Dieguez, The low temperature annealing of p-cadmium telluride in gallium-bath. Mater. Res. Bull. **28**, 1061–1066 (1993). https://doi.org/10.1016/0025-5408(93)90144-3

43. H.N. Jayatirtha, D.O. Henderson, a. Burger, M.P. Volz, Study of tellurium precipitates in CdTe crystals. Appl. Phys. Lett. **62**, 573–575 (1993). https://doi.org/10.1063/1.108885

44. S. Mazzamuto, L. Vaillant, A. Bosio, N. Romeo, N. Armani, G. Salviati, A study of the CdTe treatment with a Freon gas such as CHF2Cl. Thin Solid Films **516**, 7079–7083 (2008). https://doi.org/10.1016/j.tsf.2007.12.124

45. A. Romeo, S. Buecheler, M. Giarola, G. Mariotto, A.N. Tiwari, N. Romeo, A. Bosio, S. Mazzamuto, Study of CSS- and HVE-CdTe by different recrystallization processes. Thin Solid Films **517**, 2132–2135 (2009). https://doi.org/10.1016/j.tsf.2008.10.129

46. N.B. Chaure, A.P. Samantilleke, I.M. Dharmadasa, The effects of inclusion of iodine in CdTe thin films on material properties and solar cell performance. Sol Energy Mater Sol Cells **77**, 303–317 (2003). https://doi.org/10.1016/S0927-0248(02)00351-3

47. M.A. Green, K. Emery, Y. Hishikawa, W. Warta, E.D. Dunlop, Solar cell efficiency tables (version 46). Prog Photovolt Res Appl **23**, 805–812 (2015). https://doi.org/10.1002/pip.2637

48. First Solar raises bar for CdTe with 21.5% efficiency record: pv-magazine, (n.d.), http://www.pv-magazine.com/news/details/beitrag/first-solar-raises-bar-for-cdte-with-215-efficiency-record_100018069/#axzz3rzMESjUl. Accessed 20 Nov 2015

49. A. Rios-Flores, O. Arés, J.M. Camacho, V. Rejon, J.L. Peña, Procedure to obtain higher than 14% efficient thin film CdS/CdTe solar cells activated with HCF 2Cl gas. Sol Energy **86**, 780–785 (2012). https://doi.org/10.1016/j.solener.2011.12.002

50. O.K. Echendu, I.M. Dharmadasa, The effect on CdS/CdTe solar cell conversion efficiency of the presence of fluorine in the usual CdCl2 treatment of CdTe. Mater. Chem. Phys. **157**, 39–44 (2015). https://doi.org/10.1016/j.matchemphys.2015.03.010

51. A.A. Ojo, I.M. Dharmadasa, Electrodeposition of fluorine-doped cadmium telluride for application in photovoltaic device fabrication. Mater Res Innov **19**, 470–476 (2015). https://doi.org/10.1080/14328917.2015.1109215

52. R.J. Deokate, S.M. Pawar, a.V. Moholkar, V.S. Sawant, C.a. Pawar, C.H. Bhosale, K.Y. Rajpure, Spray deposition of highly transparent fluorine doped cadmium oxide thin films. Appl. Surf. Sci. **254**, 2187–2195 (2008). https://doi.org/10.1016/j.apsusc.2007.09.006

53. P. Ghosh, Electrical and optical properties of highly conducting CdO:F thin film deposited by sol–gel dip coating technique. Sol. Energy Mater. Sol. Cells. **81**, 279–289 (2004). https://doi.org/10.1016/j.solmat.2003.11.021

54. E. Elangovan, K. Ramamurthi, Studies on micro-structural and electrical properties of spray-deposited fluorine-doped tin oxide thin films from low-cost precursor. Thin Solid Films **476**, 231–236 (2005). https://doi.org/10.1016/j.tsf.2004.09.022

55. V.D. Popovych, I.S. Virt, F.F. Sizov, V.V. Tetyorkin, Z.F. Tsybrii (Ivasiv), L.O. Darchuk, O.A. Parfenjuk, M.I. Ilashchuk, The effect of chlorine doping concentration on the quality of

CdTe single crystals grown by the modified physical vapor transport method. J. Cryst. Growth **308**, 63–70 (2007). https://doi.org/10.1016/j.jcrysgro.2007.07.041

56. I.M. Dharmadasa, Recent developments and progress on electrical contacts to CdTe, CdS and ZnSe with special reference to barrier contacts to CdTe. Prog Cryst Growth Charact Mater **36**, 249–290 (1998). https://doi.org/10.1016/S0960-8974(98)00010-2

57. L. Jin, Y. Linyu, J. Jikang, Z. Hua, S. Yanfei, Effects of Sn-doping on morphology and optical properties of CdTe polycrystalline films. J Semicond **30**, 112003 (2009). https://doi.org/10.1088/1674-4926/30/11/112003

58. M. Kul, A.S. Aybek, E. Turan, M. Zor, S. Irmak, Effects of fluorine doping on the structural properties of the CdO films deposited by ultrasonic spray pyrolysis. Sol. Energy Mater. Sol. Cells. **91**, 1927–1933 (2007). https://doi.org/10.1016/j.solmat.2007.07.014

59. A. Monshi, Modified Scherrer equation to estimate more accurately nano-crystallite size using XRD. World J Nano Sci Eng **2**, 154–160 (2012). https://doi.org/10.4236/wjnse.2012.23020

60. B.G. Yacobi, *Semiconductor Materials: An Introduction to Basic Principles*, 1st edn. (Springer, New York, 2003). https://doi.org/10.1007/b105378

61. T. Schulmeyer, J. Fritsche, A. Thißen, A. Klein, W. Jaegermann, M. Campo, J. Beier, Effect of in situ UHV CdCl2-activation on the electronic properties of CdTe thin film solar cells. Thin Solid Films **431–432**, 84–89 (2003). https://doi.org/10.1016/S0040-6090(03)00207-4

62. I.M. Dharmadasa, P. Bingham, O.K. Echendu, H.I. Salim, T. Druffel, R. Dharmadasa, G. Sumanasekera, R. Dharmasena, M.B. Dergacheva, K. Mit, K. Urazov, L. Bowen, M. Walls, A. Abbas, Fabrication of CdS/CdTe-based thin film solar cells using an electrochemical technique. Coatings. **4**, 380–415 (2014). https://doi.org/10.3390/coatings4030380

63. I.M. Dharmadasa, W.G. Herrenden-Harker, R.H. Williams, Metals on cadmium telluride: Schottky barriers and interface reactions. Appl. Phys. Lett. **48**, 1802 (1986). https://doi.org/10.1063/1.96792

64. J.M. Burst, J.N. Duenow, D.S. Albin, E. Colegrove, M.O. Reese, J.A. Aguiar, C.-S. Jiang, M.K. Patel, M.M. Al-Jassim, D. Kuciauskas, S. Swain, T. Ablekim, K.G. Lynn, W.K. Metzger, CdTe solar cells with open-circuit voltage breaking the 1 V barrier. Nat Energy **1**, 16015 (2016). https://doi.org/10.1038/nenergy.2016.15

65. M.O. Reese, C.L. Perkins, J.M. Burst, S. Farrell, T.M. Barnes, S.W. Johnston, D. Kuciauskas, T.A. Gessert, W.K. Metzger, Intrinsic surface passivation of CdTe. J. Appl. Phys. **118**, 155305 (2015). https://doi.org/10.1063/1.4933186

66. D.G. Diso, F. Fauzi, O.K. Echendu, O.I. Olusola, I.M. Dharmadasa, Optimisation of CdTe electrodeposition voltage for development of CdS/CdTe solar cells. J. Mater. Sci. Mater. Electron. **27**, 12464–12472 (2016). https://doi.org/10.1007/s10854-016-4844-3

67. M.K. Rabinal, I. Lyubomirsky, E. Pekarskaya, V. Lyakhovitskaya, D. Cahen, Low resistance contacts to p-CuInSe2 and p-CdTe crystals. J. Electron. Mater. **26**, 893–897 (1997). https://doi.org/10.1007/s11664-997-0270-x

68. S. Nozaki, A.G. Milnes, Specific contact resistivity of indium contacts to n-type CdTe. J. Electron. Mater. **14**, 137–155 (1985). https://doi.org/10.1007/BF02656672

69. I.M. Dharmadasa, A.P. Samantilleke, N.B. Chaure, J. Young, New ways of developing glass/conducting glass/CdS/CdTe/metal thin-film solar cells based on a new model. Semicond. Sci. Technol. **17**, 1238–1248 (2002). https://doi.org/10.1088/0268-1242/17/12/306

70. I.M. Dharmadasa, C.J. Blomfield, C.G. Scott, R. Coratger, F. Ajustron, J. Beauvillain, Metal/n-CdTe interfaces: a study of electrical contacts by deep level transient spectroscopy and ballistic electron emission microscopy. Solid State Electron. **42**, 595–604 (1998). https://doi.org/10.1016/S0038-1101(97)00296-7

71. H. Metin, R. Esen, Annealing effects on optical and crystallographic properties of CBD grown CdS films. Semicond Sci Technol **18**, 647–654 (2003). https://doi.org/10.1088/0268-1242/18/7/308

72. I.M. Dharmadasa, J.D. Bunning, A.P. Samantilleke, T. Shen, Effects of multi-defects at metal/semiconductor interfaces on electrical properties and their influence on stability and lifetime of thin film solar cells. Sol. Energy Mater. Sol. Cells. **86**, 373–384 (2005). https://doi.org/10.1016/j.solmat.2004.08.009

73. R. Paterson, J. Anderson, S.S. Anderson, Lutfullah, Transport in aqueous solutions of group IIB metal salts at 298.15 K. Part 2.—Interpretation and prediction of transport in dilute solutions of cadmium iodide: an irreversible thermodynamic analysis. J. Chem. Soc. Faraday Trans. 1 Phys. Chem. Condens. Phases. **73**, 1773 (1977). https://doi.org/10.1039/f19777301773

74. R. Paterson, C. Devine, Transport in aqueous solutions of Group IIB metal salts. Part 7.— Measurement and prediction of isotopic diffusion coefficients for iodide in solutions of cadmium iodide. J Chem Soc Faraday Trans 1 Phys Chem Condens Phases **76**, 1052 (1980). https://doi.org/10.1039/f19807601052

75. A. Bouraiou, M.S. Aida, O. Meglali, N. Attaf, Potential effect on the properties of CuInSe2 thin films deposited using two-electrode system. Curr Appl Phys **11**, 1173–1178 (2011). https://doi.org/10.1016/j.cap.2011.02.014

76. C. Ferekides, J. Britt, CdTe solar cells with efficiencies over 15%. Sol. Energy Mater. Sol. Cells. **35**, 255–262 (1994). https://doi.org/10.1016/0927-0248(94)90148-1

77. A.A. Ojo, I.M. Dharmadasa, The effect of fluorine doping on the characteristic behaviour of CdTe. J. Electron. Mater. **45**, 5728–5738 (2016). https://doi.org/10.1007/s11664-016-4786-9

78. M.A. Redwan, E.H. Aly, L.I. Soliman, A.A. El-Shazely, H.A. Zayed, Characteristics of n-Cd0.9 Zn0.1S/p-CdTe heterojunctions. Vacuum **69**, 545–555 (2003). https://doi.org/10.1016/S0042-207X(02)00604-8

79. B.M. Basol, Electrodeposited CdTe and HgCdTe solar cells. Sol Cells **23**, 69–88 (1988). https://doi.org/10.1016/0379-6787(88)90008-7

80. T.M. Razykov, N. Amin, B. Ergashev, C.S. Ferekides, D.Y. Goswami, M.K. Hakkulov, K.M. Kouchkarov, K. Sopian, M.Y. Sulaiman, M. Alghoul, H.S. Ullal, Effect of CdCl2 treatment on physical properties of CdTe films with different compositions fabricated by chemical molecular beam deposition. Appl Sol Energy **49**, 35–39 (2013). https://doi.org/10.3103/S0003701X1301009X

81. P.J. Sellin, A.W. Dazvies, A. Lohstroh, M.E. Özsan, J. Parkin, Drift mobility and mobility-lifetime products in CdTe:Cl grown by the travelling heater method. IEEE Trans. Nucl. Sci. **52**, 3074–3078 (2005). https://doi.org/10.1109/TNS.2005.855641

82. M.A. Cengel, Y.A. Boles, *Thermodynamics: An Engineering approach*, 5th edn. (McGraw-Hill, New York, 2006)

83. V.P. Vasil'ev, Correlations between the thermodynamic properties of II--VI and III--VI phases. Inorg Mater **43**, 115–124 (2007). https://doi.org/10.1134/S0020168507020045

84. O. Chang-Seok, L.D. Nyung, C. Oh, Thermodynamic assessment of the Ga·Te system. Calphad **16**, 317–330 (1992). https://doi.org/10.1016/0364-5916(92)90029-W

85. J. Han, C. Spanheimer, G. Haindl, G. Fu, V. Krishnakumar, J. Schaffner, C. Fan, K. Zhao, a. Klein, W. Jaegermann, Optimized chemical bath deposited CdS layers for the improvement of CdTe solar cells. Sol. Energy Mater. Sol. Cells. **95**, 816–820 (2011). https://doi.org/10.1016/j.solmat.2010.10.027

86. E. PŁaczek-Popko, Z. Gumienny, J. Trzmiel, J. Szatkowski, Evidence for metastable behavior of Ga-doped CdTe. Opt Appl **38**, 559–565 (2008)

87. K.C. Mandal, R.M. Krishna, T.C. Hayes, P.G. Muzykov, S. Das, T.S. Sudarshan, S. Ma, Layered GaTe crystals for radiation detectors. IEEE Trans. Nucl. Sci. **58**, 1981–1986 (2011). https://doi.org/10.1109/TNS.2011.2140330

88. A.M. Mancini, C. Manfredotti, A. Rizzo, G. Micocci, Vapour growth of GaTe single crystals. J. Cryst. Growth **21**, 187–190 (1974). https://doi.org/10.1016/0022-0248(74)90003-7

Chapter 7
Solar Cell Fabrication and Characterisation

7.1 Introduction

This chapter presents the characterisation of solar cell devices fabricated using the pre-characterised ZnS, CdS and CdTe layers as documented in Chaps. 3, 4 and 5. For the fabricated PV devices as presented in this book, ZnS was used as the buffer layer, CdS is utilised as the window layer, while CdTe is utilised as the main absorber layer in forming the solar cell structure. Thus, the main solar cell structure explored incorporates a CdS/CdTe heterojunction core. This chapter systematically reports the effect of the incorporation of a buffer layer to base CdS/CdTe configuration and the effect of various window layer conditions on the device properties of PV devices. This is followed by the exploration of the effect of different conditions of CdTe absorber layer and post-growth treatment of the device properties. Further to this, the effect of the extrinsic doping of CdTe, metal contacts, various heat treatment temperatures, etching and the incorporation of pinhole plugin layers into the CdS-/CdTe-based PV devices was also discussed (see Fig. 1.6).

7.2 Basic Solar Cell Fabrication Process: Post-growth Treatment, Etching Process and Device Fabrication of CdS-/CdTe-Based Photovoltaic Devices

In this study, two aqueous solutions (A and B) were used to treat the material layers. Solution A contains 0.1 M $CdCl_2$ in 20 mL of DI water at room temperature, while solution B contains 0.1 M $CdCl_2$ and 0.05 M $Ga_2(SO_4)_3$ in 20 mL of DI water at room temperature. The solutions were stirred continuously for 1 h to achieve homogeneity. The sample labelled AD was left as-deposited, while samples marked CCT and GCT were $CdCl_2$ treated in solution A and $CdCl_2$/Ga treated in solution B, respectively.

© Springer International Publishing AG, part of Springer Nature 2019
A. A. Ojo et al., *Next Generation Multilayer Graded Bandgap Solar Cells*,
https://doi.org/10.1007/978-3-319-96667-0_7

The application of $CdCl_2$ or $CdCl_2 + Ga_2(SO_4)_3$ on the grown CdTe single layers (such as glass/FTO/CdS/CdTe during material characterisation) or multilayer structure (such as glass/FTO/n-CdS/n-CdTe/p-CdTe during device fabrication) was achieved by adding a few drops of relevant solution on their surface. The full coverage of the layers with the treatment solutions was achieved by spreading the solution using solution-damped cotton buds. Both the CCT- and GCT-treated glass/FTO/n-CdS/n-CdTe/p-CdTe layers were allowed to air-dry before heat treatment.

The post-growth heat treatments were performed within the range of 350–450 °C for 60–20 min in an air atmosphere for samples undergoing each treatment based on previously optimised conditions [1–4]. The CCT- and GCT-treated layers were then rinsed in DI water and dried in a stream of nitrogen afterwards. It should be noted that the motivation for the incorporation of halides such as chlorine, fluorine and other elements such as gallium in the post-growth treatment has been well documented in the literature [1, 5, 6] and in Chap. 6.

Characterisation of layers proceeds immediately for single semiconductor layers treated to explore their material property. While for device-bound structures, the surfaces of the AD, CCT and GCT layers were etched using a solution containing $K_2Cr_2O_7$ and concentrated H_2SO_4 for acid etching and a solution containing NaOH and $Na_2S_2O_3$ for alkalline etching for 2 s and 2 min, respectively, to improve the metal/semiconductor contact [7, 8]. Immediately afterwards, the AD, CCT and GCT samples were transferred to a high vacuum system in order to deposit 2 mm diameter and 100-nm-thick Au contacts on the device structure. The fabricated devices were analysed using characteristic device measurements (see Sect. 3.5) to determine their device parameters.

7.3 Effect of CdS Thickness in Glass/FTO/n-CdS/n-CdTe/ p-CdTe/Au Devices

Further to the effect of CdS thickness on its material properties as discussed in Sect. 5.4, glass/FTO/n-CdS/n-CdTe/p-CdTe/Au solar cell devices were fabricated using different CdS thicknesses. 1200-nm-thick n-CdTe layers were electrodeposited at 1370 mV, followed by the deposition of ~30 nm p-CdTe at 1360 mV (see Sect. 6.3.8) in a continuous deposition process to achieve glass/FTO/n-CdS/n-CdTe/ p-CdTe configuration.

Figure 7.1a shows the band diagram of the glass/FTO/n-CdS/n-CdTe/p-CdTe/Au device, and Fig. 7.1b shows the I-V curve of a cell incorporating 150-nm-thick CdS layer at AM1.5 condition, while Fig. 7.1c, d show the linear-linear and log-linear I-V curves of the glass/FTO/n-CdS/n-CdTe/p-CdTe/Au devices incorporating 150 nm CdS layer under dark condition.

The band diagram as depicted in Fig. 7.1a is a result of prior material investigation as reported in the literature by Salim et al. [9] that the conductivity type of

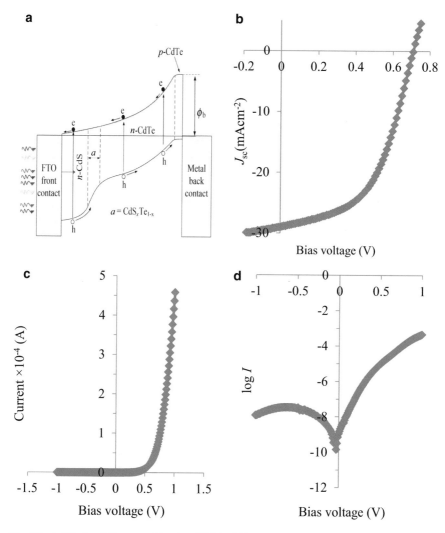

Fig. 7.1 (**a**) The band diagram of the glass/FTO/*n*-CdS/*n*-CdTe/*p*-CdTe/Au thin-film solar cell and (**b**) current-voltage curve of glass/FTO/*n*-CdS/*n*-CdTe/*p*-CdTe/Au champion cell incorporating 150 nm CdS, while (**c**) and (**d**) are the linear-linear and log-linear *I-V* curves under dark condition

electrodeposited CdTe layer is retained after $CdCl_2$ treatment, although a shift towards the opposite conductivity type was observed.

Figures 7.2a–d show the short-circuit current density (J_{sc}), open-circuit voltage (V_{oc}), fill factor (FF) and solar energy conversion efficiency (η) measured at AM1.5 against CdS layer thickness, respectively, while Table 7.1 summarises the tabulated *I-V* parameters of the three champion cells from the solar devices with varied CdS thickness.

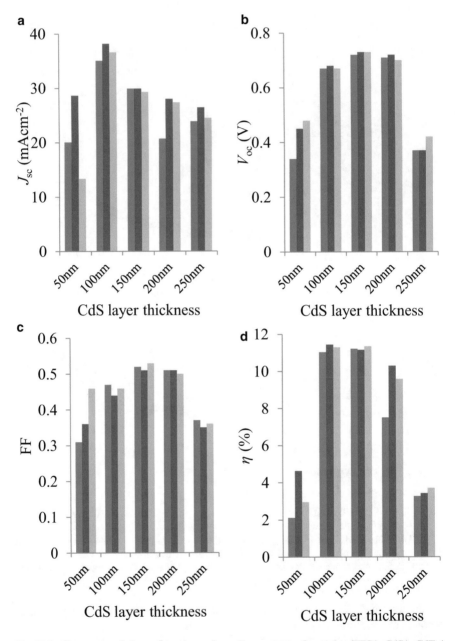

Fig. 7.2 Histograms of three champion solar cell parameters from glass/FTO/n-CdS/n-CdTe/ p-CdTe/Au with different CdS thicknesses measured at AM1.5 (**a**) J_{sc}, (**b**) V_{oc}, (**c**) FF and (**d**) η, against CdS thickness

Table 7.1 Tabulated device parameters for three champion cells of glass/FTO/*n*-CdS/*n*-CdTe/*p*-CdTe/Au incorporating different CdS thicknesses

CdS (nm)	J_{sc} (mAcm^{-2})	V_{oc} (mV)	FF	Efficiency (%)
50	20.1	0.34	0.31	2.11
	28.7	0.45	0.36	4.64
	13.4	0.48	0.46	2.95
100	35.0	0.67	0.47	11.03
	38.2	0.68	0.44	11.43
	36.6	0.67	0.46	11.29
150	29.9	0.72	0.52	11.29
	29.9	0.73	0.51	11.15
	29.3	0.73	0.53	11.34
200	20.7	0.71	0.51	7.50
	28.0	0.72	0.51	10.29
	27.4	0.70	0.50	9.59
250	23.9	0.37	0.37	3.27
	26.4	0.37	0.35	3.42
	24.5	0.42	0.36	3.71

As shown in Fig. 7.2a, the comparatively low J_{sc} observed in the device incorporating the 50 nm CdS layer was not expected due to the high transmittance in 50-nm-thick CdS layer as discussed in Sect. 5.4.2. Hence, the high photocurrent is expected. But based on the incomplete coverage of the underlying glass/FTO substrate, the gaps will serve as shunting paths and thereby reducing the photo-generated current due to increased recombination of the electron-hole pairs and poor-quality diode.

The increase in the CdS thickness in the device configuration to 100 nm shows higher J_{sc} due to better CdS layer coverage over the glass/FTO and high transmittance as discussed in Sect. 5.4.2. The devices incorporating CdS with thickness above 100 nm show a gradual reduction in J_{sc}. This might be as a result of the reduction in transmittance, increase in series resistance and also increased parasitic absorption due to increased CdS thickness [5, 10]. Granata et al. demonstrated low CdS thickness (40 nm) [10] to achieve the highest photocurrent using high-temperature (600 °C) closed-space sublimation growth technique due to the unique qualities of CdS. However, electrodeposition at low temperature (~85 °C), on the other hand, requires an increased thickness of CdS layer to suppress the effect of surface roughness and deposition/nucleation mechanism. It should be noted that the J_{sc} as observed in this work is higher than the Shockley-Queisser limit of ~26 mAcm^{-2} on single *p-n* junction incorporating CdTe absorber layer [11] due to the incorporation of the multilayer *n-n-p* configuration [12].

An increase in both V_{oc} and FF with increasing CdS thickness up to 200 nm as shown in Fig. 7.2b, c is also observed. This observation can be associated with the reduction of shunts [13] and other defects relating to the early CdS nucleation stages. Ultimately, the efficiency of devices incorporating both the 100- and 150-nm-thick

Table 7.2 Diode parameters extracted from dark I-V for champion cells of glass/FTO/n-CdS/n-CdTe/p-CdTe/Au incorporating different CdS thicknesses

CdS (nm)	R_{sh} (Ω)	R_s (kΩ)	log (RF)	I_o (A)	n	ϕ_b (eV)
50	51.1	0.02	0.3	2.81×10^{-3}	>2.00	>0.42
100	4.3×10^6	0.50	4.1	1.02×10^{-9}	1.86	>0.80
150	6.2×10^6	0.57	4.4	1.26×10^{-9}	1.71	>0.80
200	3.5×10^6	0.80	3.5	3.98×10^{-9}	1.95	>0.77

CdS layer appears to be the highest with fairly similar efficiency values due to comparatively high J_{sc} observed in the 100 nm CdS and high V_{oc} plus FF in the device incorporating 150-nm-thick CdS layer. It should be taken into account that shunt resistance is associated with semiconductor layer quality [14] which explains the low efficiency observed in the device incorporating the electrodeposited 50-nm-thick CdS layer. The gradual reduction in efficiency with increasing thickness is due to the low transmittance and reduced photocurrent as a result of increased parasitic absorption [13].

Table 7.2 shows the diode parameters such as shunt resistance R_{sh}, series resistance R_s, rectification factor RF, saturated current I_o, ideality factor n and barrier height ϕ_b as obtained from the champion cells of the glass/FTO/n-CdS/n-CdTe/p-CdTe/Au devices incorporating different CdS- thicknesses tabulated in Table 7.1. As observed in Table 7.2, the low R_{sh} observed for devices incorporating 50-nm-thick CdS layer signifies the presence of shunt paths. The shunting might be due to the incomplete coverage of the glass/FTO with CdS as discussed in Sect. 5.4, thereby creating direct leakage paths between the glass/FTO substrate and the grown CdTe layers as further suggested by the comparatively high I_o value observed using the 50-nm-thick CdS layer. From observation, the R_s under dark condition is high for layers incorporating CdS thickness of 100 nm and above. However, I-V measurements under illuminated condition show a reduction in R_s value to 100–200 Ω range. This reduction is due to the photoconductivity of the material layer used. As iterated by Dharmadasa [15], high-efficiency solar cells can only be achieved provided the RF value is $\geq 10^3$. The RF values as observed for devices incorporating CdS with thickness ≥ 100 nm show the tendency of achieving high efficiency. Furthermore, the ideality factor n which depicts the charge carrier transportation mechanisms shows that the glass/FTO/n-CdS/n-CdTe/p-CdTe/Au devices incorporating CdS with a thickness between 100 and 200 nm are governed by both recombination and generation (R&G) centres and thermionic emission. The devices with the 50 nm CdS layers have the highest n values ($n > 2.00$). This is due to low-quality CdTe material deposited on FTO regions instead of the fully covered FTO layer with CdS. CdS provides a high-quality substrate for growing superior CdTe layers with low defect densities. The lowest ideality factors observed for 100 and 150 nm show the presence of low defects reducing R&G process. Therefore the J_{sc} values can be large for such devices as we experimentally observed in this work.

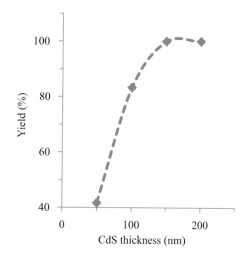

Fig. 7.3 Graph of percentage yield against CdS thickness in a glass/FTO/*n*-CdS/*n*-CdTe/*p*-CdTe/Au device structure

7.3.1 Photovoltaic (PV) Device Yield

Figure 7.3 shows the graph of the percentage yield of the active cells as against the CdS thickness incorporated in the glass/FTO/*n*-CdS/*n*-CdTe/*p*-CdTe/Au structure. The number of cells in all the fabricated solar cell devices incorporating different CdS window layer thicknesses was 12. It was discovered that the percentage of PV-active cell in the device incorporating 50-nm-thick CdS was low. This observation can be related to the morphological analysis as discussed in Sect. 5.4.3.

There is a high tendency that the non-functional cells are glass/FTO/*n*-CdTe/*p*-CdTe/Au due to the incomplete coverage of the underlying CdS layer. It should be noted that although a depletion layer is expected in between the *n*-CdTe and *p*-CdTe, due to the characteristic large grains observed in CdTe after treatment, the increased grain growth of CdTe on a rough surface favours the inclusion of pores or pinholes in between the grains after treatment. This leaves Au with the tendency of contacting directly with glass/FTO creating shunting paths and electron-hole pair recombination centres. An increase in the CdS thickness above 50 nm improves the percentage yield close to 100% at 150 nm CdS thickness.

7.3.2 Standard Deviation

Figure 7.4 shows the normal distribution curve of both J_{sc} and η using standard deviation parameters as presented in Table 7.3. As stated in Sect. 7.3.1, only three champion cells from each fabricated device incorporating 50-, 100- and 150-nm-thick CdS were considered in this work due to high parasitic absorption observed with higher CdS thickness. Both the mean ($\bar{x}$) and the standard deviation (δ) were

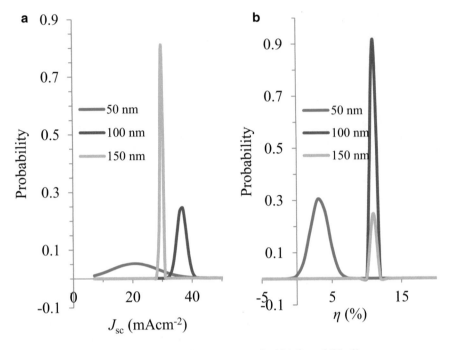

Fig. 7.4 Normal curves of probability of occurrence against (**a**) J_{sc} and (**b**) $\eta\%$

Table 7.3 Table of mean and standard deviation of three champion cell device parameters incorporating different CdS thicknesses in glass/FTO/n-CdS/n-CdTe/p-CdTe/Au configuration

CdS (nm)	Mean $\bar{x} = \dfrac{1}{n}\sum\limits_{i=1}^{n} x_i$				Standard deviation $\delta = \sqrt{\dfrac{\Sigma(x-\bar{x})^2}{n-1}}$			
	J_{sc} (mAcm^{-2})	V_{oc} (V)	Fill factor	η (%)	J_{sc} (mAcm^{-2})	V_{oc} (V)	Fill factor	η (%)
50	20.70	0.423	0.38	3.24	7.66	0.074	0.0764	1.2879
100	36.62	0.673	0.46	11.25	1.59	0.006	0.0153	0.2039
150	29.72	0.726	0.52	11.23	0.37	0.006	0.0100	0.0971
200	25.37	0.710	0.51	9.12	4.06	0.010	0.0058	1.4537
250	24.95	0.386	0.36	3.47	1.33	0.029	0.0100	0.2222

calculated as shown in Equations 7.1 and 7.2, respectively, where n is the number of cells explored and x is the value of the parameter of the explored cells.

$$\bar{x} = \frac{1}{n}\sum\nolimits_{i=1}^{n} x_i \qquad \text{(Equation 7.1)}$$

$$\delta = \sqrt{\frac{\sum(x-\bar{x})^2}{n-1}} \qquad \text{(Equation 7.2)}$$

It should be noted that the data utilised in the calculations shown in Table 7.3 is an extract from Table 7.2. The standard deviation as depicted using the normal distribution curve is used to measure the dispersion in a set of explored data. The horizontal axis represents the dependent variable which for the sake of the work is J_{sc} and η in Fig. 7.4a, b, respectively, while the vertical axis shows the probability that the value of the standard deviation will occur.

As shown in Fig. 7.4a, the highest J_{sc} was observed in the device incorporating the 100-nm-thick CdS layer with a 50% probability of achieving 36.62 mAcm^{-2} and ~30% probability of occurrence as compared to the devices incorporating 150-nm-thick CdS layer with 50% probability of achieving a J_{sc} of 29.72 mAcm^{-2} and ~80% probability of occurrence. In both the devices incorporating 100- and 150-nm-thick CdS, interesting features such as high J_{sc} and high probability of occurrence can be explored by carefully optimising the CdS layer thickness to fall in between 100 and 150 nm CdS thickness.

It should be noted that one of the advantages related to an increase in the thickness of the CdS window layer is smoothening out the surface before absorber layers such as CdTe can be grown. Devices incorporating the 50-nm-thick CdS layers show the largest dispersion of J_{sc}, lowest J_{sc} value range and less than 10% probability of achieving comparably high photocurrent based on the results obtained from this work. For example, based on the normal curve generated in Fig. 7.4a, less than 5% of the fabricated cells in a device can achieve J_{sc} of 30 mAcm^{-2} and above for devices incorporating 50 nm CdS thickness, while devices incorporating 100-nm and 150-nm-thick CdS show 100% and 50% probability of achieving 30 mAcm^{-2} short-circuit current density, respectively. The effect of both pinholes and parasitic absorption is annotated for the low (50 nm) and high (150 nm) CdS thickness in this work.

Furthermore, Fig. 7.4b shows that the highest η can be achieved with devices incorporating both 100- and 150-nm-thick CdS with the high occurrence probability of ~90% observed at 100 nm and ~30% occurrence probability for the 150-nm-thick CdS. It should be noted that the η of the device incorporating the 50-nm-thick CdS shows the low η value range and low probability of occurrence.

7.3.3 Summations

In conclusion, this work supports the work done by other researchers on the optimisation of CdS thickness as related to CdS/CdTe cell but focuses on the iteration of the effect of deposition technique and nucleation mechanism of electroplated semiconductor materials. But contrarily, the proposed thin CdS window layer with thickness <50 nm by other authors cannot achieve comparatively high efficiency using electrodeposition technique without detrimental effect as demonstrated in this section.

From the observed results, optical properties such as transmittance favour the 50 nm CdS thickness, while crystallinity favoured the highest CdS thickness with increase in both crystallite size and preferred orientation reflection intensity. But with a more critical observation on PV property and statistical analysis, the glass/FTO/n-CdS/n-CdTe/p-CdTe/Au shows better device property with the incorporation of 100–150-nm-thick CdS layer relative to the optical, morphological, structural and electronic properties analyses of the incorporated CdS layer.

7.4   Effect of CdTe Growth Voltage on the Efficiency of a Simple CdS-/CdTe-Based Solar Cell

Further to the summations made on the effect of Cd/Te atomic composition in the determination of the conduction type of CdTe (see compositional and PEC analysis in Sects. 5.3.7 and 5.3.8), CdTe layers were grown on similar glass/FTO/n-CdS layers. The ~120-nm-thick CdS layers incorporated were grown on a 6×5 cm^2 glass/FTO layer. The electrolytic bath setup was in accordance with the CdS growth parameters as discussed in Sect. 5.2.1. The CdS layers were CdCl$_2$ treated at 400 °C for 20 min in the air. Prior to the deposition of CdTe, the glass/FTO/n-CdS layers were air-cooled and rinsed in DI water and cut into six 1×5 cm^2 glass/FTO/n-CdS strips. ~1200-nm-thick CdTe layers were grown at cathodic voltages between 1340 and 1400 mV (see Fig. 7.5) on the individual glass/FTO/n-CdS layers to give either glass/FTO/n-CdS/p-CdTe/Au (for $V_g < V_i$) or glass/FTO/n-CdS/n-CdTe/Au (for $V_g > V_i$) layers after metallisation. The band diagrams of both device configurations are shown in Fig. 7.6, the I-V curves are shown in Fig. 7.7, while the measured parameters are tabulated in Table 7.4.

It should be noted that due to the Fermi level (FL) pinning at the n-CdTe/metal interface for the device configuration as shown in Fig. 7.6b [7, 16] a large Schottky barrier (SB) can be formed at this interface. Consequently, the CdS/CdTe/metal structure forms an n-n + SB structure with two PV-active junctions. The photo-generated current components produced by these two rectifying interfaces (n-n + SB) add up together; therefore, the two junctions are connected in parallel. Consequently, this configuration can be classified as a tandem solar cell with two junctions connected in parallel [17, 18] and capable of producing improved performance such as J_{sc} above the Shockley-Queisser limit [11] on single p-n junction device.

The narrative of the effect of near-stoichiometric, Te-rich and Cd-rich can be better described using Table 7.4 which is further to the visual, optical and photoelectrochemical observations as shown in Fig. 7.5. Due to brevity, a few crucial points will be focused on. The low R_{sh} as observed with devices incorporating p-CdTe grown at 1340 mV ($\sim$−30 mV away from V_i) is an indication of low material quality as inferred in the literature [19] and attributed to the presence of Te precipitates, voids, gaps and high dislocation density within the semiconductor material.

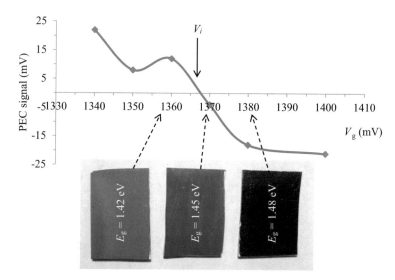

Fig. 7.5 The physical appearance of CdTe layers grown close to the transition voltage (V_i = 1.368 V). Colour of the CdTe layer varies from dark, light-dark to honey colour and transparent when moved from Te richness to Cd richness. The average bandgap values of five measurements are also indicated for layers grown in this region

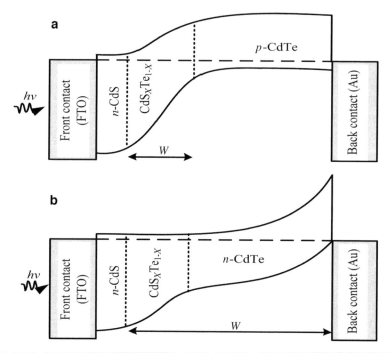

Fig. 7.6 Energy band diagrams representing (**a**) glass/FTO/n-CdS/p-CdTe/Au and (**b**) glass/FTO/ n-CdS/n-CdTe/Au device configurations

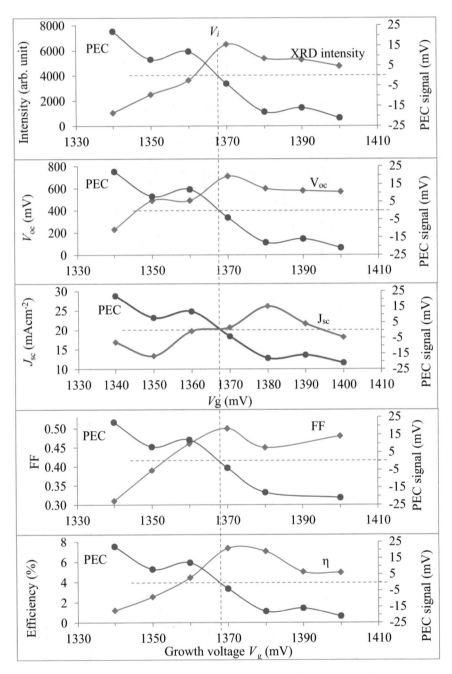

Fig. 7.7 Typical variation of the PEC signal for CdCl$_2$-treated CdTe is shown in each frame to separate the *p*-type and *n*-type regions. The intensity of (111) XRD peak is shown in the first frame indicating best crystallinity around V_i. The device parameters for glass/FTO/CdS/CdTe/Au solar cells fabricated with the CdTe layers grown in the vicinity of the transition voltage, $V_i = 1368$ mV, are shown in the other four frames

Table 7.4 Summary of device parameters obtained from *I-V* (both under illuminated and dark conditions) and *C-V* (dark condition) for simple CdS-/CdTe-based solar cells grown at different growth voltages in the vicinity of V_i

CdTe growth voltage (mV)	1340	1360	1370	1380	1400
I-V under dark condition					
R_{sh} (Ω)	1016	$>10^5$	$>10^5$	$>10^5$	$>10^5$
R_s (kΩ)	0.21	0.80	0.50	1.43	1.50
log (RF)	0.4	3.5	3.9	3.3	3.0
I_o (A)	2.5×10^{-5}	3.9×10^{-9}	1.0×10^{-9}	3.2×10^{-9}	5.0×10^{-9}
n	>2.00	1.95	1.86	1.58	1.86
ϕ_b (eV)	>0.52	>0.76	>0.81	>0.77	>0.77
I-V under AM1.5 illumination condition					
I_{sc} (mA)	0.53	0.62	0.65	0.82	0.57
J_{sc} (mAcm^{-2})	16.88	19.75	20.70	26.11	18.15
V_{oc} (V)	0.23	0.49	0.72	0.60	0.57
Fill factor	0.31	0.46	0.50	0.45	0.48
Efficiency (%)	1.20	4.45	7.50	7.05	4.97
C-V under dark condition					
$\sigma \times 10^{-4}$ (Ω.cm)$^{-1}$	1.41	–	2.85	–	6.03
N_D (cm^{-3})	7.74×10^{16}	–	3.10×10^{14}	–	9.10×10^{14}
μ (cm^{2}V^{-1} s^{-1})	0.01	–	5.74	–	4.14
C_o (pF)	1630	–	330	–	370
W (nm)	187.6	–	926.7	–	826.5

On the other hand, slight reductions in R_{sh} were recorded for the devices incorporating *n*-CdTe grown at 1400 mV (~+30 mV away from V_i). Further observation of parameters including the RF, I_o, n and ϕ_b of the fabricated devices measured using *I-V* curves under dark condition shows comparatively preferred electronic property for devices incorporating *n*-CdTe absorber layers. Taking into consideration that log(*RF*) of 3 is a requirement of high-efficiency solar cells [15] and the observed *n* values were within 1.00 and 2.00, this signifies that the current transport mechanisms consist of both thermionic emission and R&G for devices incorporating CdTe layers grown between 1360 and 1400 mV. The *n*-value of devices incorporating *p*-CdTe grown at 1340 mV (*n* > 2.00) indicates that the Te-rich CdTe layers have more defects contributing to R&G process. This observation may be the probable cause of the reduction in the barrier height ϕ_b, while the comparatively low depletion width of 187.6 nm (as shown in Table 7.4) may have aided additional tunnelling in the fabricated devices as discussed in Sect. 2.3.4. It is therefore not surprising that the resulting *I-V* parameters measured under AM1.5 illuminated condition for 1340 mV incorporating *p*-CdTe are the lowest. This may either be due to the incorporation of midgap defects due to superfluous Te or sublimation of Te precipitates during post-growth heat treatment increasing pinhole density. It should be noted that the fabricated devices incorporating *p*-CdTe grown at 1360 mV, which is possibly at ppm or ppb level doping of CdTe with Te, show high conversion efficiency as compared to the 1340 mV p-CdTe device due to the growth of CdTe at the vicinity of V_i. On the other hand, devices incorporating excess Cd within the explored cathodic voltage of 1370–1400 mV show minimal parameter alteration with the best conversion efficiency obtained with devices incorporating n-CdTe grown at 1370 mV. It should be noted that the inclusion of Te precipitates in the growth of CdTe notwithstanding the technique of growth is the main detriment of CdTe [6].

7.5   Comparative Analysis of n-CdS/n-CdTe and n-ZnS/n-CdS/n-CdTe Devices

The incorporation of thin n-ZnS as a buffer layer into the n-CdS/n-CdTe device configuration (see Sect. 7.4) is mainly due to its wetting ability [20] and optoelectronic properties. ZnS with a bandgap of ~3.70 eV [21] shows high capability as a buffer layer due to its high transparency in the short wavelength region (350–550 nm) as compared to CdS with a bandgap of 2.42 eV. The 50-nm-thick ZnS incorporated in the glass/FTO/n-ZnS/n-CdS/n-CdTe/Au device configuration was electroplated from an aqueous electrolytic bath containing 0.2 M ZnSO$_4 \cdot$H$_2$O and 0.2 M (NH$_4$)$_2$S$_2$O$_3$ as Zn and S precursor, respectively [22]. The full description of the growth and characterisation of ZnS layers are described in Chap. 4. Prior to the deposition of CdS, the n-ZnS was heat-treated at 300 °C and rinsed after being air-cooled. The procedure for the deposition of the n-CdS/n-CdTe layers for both device architectures under comparison is similar. The thicknesses of both glass/FTO/n-CdS/n-CdTe/Au and glass/FTO/n-ZnS/n-CdS/n-CdTe/Au prior to any treatment is glass/FTO/120 nm/1200 nm/100 nm and glass/FTO/50 nm/65 nm/1200 nm/100 nm, respectively.

The band diagram of the glass/FTO/n-CdS/n-CdTe/Au is similar to Fig. 7.6b, while the glass/FTO/n-ZnS/n-CdS/n-CdTe/Au configuration is shown in Fig. 7.8. The presence of interfacial ternary compounds such as Zn$_x$Cd$_{1-x}$S and CdS$_x$Te$_{1-x}$ was incorporated in Fig. 7.8, as it is well documented in the literature that the formation of Zn$_x$Cd$_{1-x}$S is due to the interdiffusion of Zn and Cd between the ZnS and the CdS, while the formation of CdS$_x$Te$_{1-x}$ is due to the interdiffusion of S and Te between the CdS and CdTe layers during post-growth annealing in the presence of CdCl$_2$ [23–25].

Apart from the advantages of graded bandgap configuration such as harnessing photons from UV, visible and IR from the electromagnetic spectrum as discussed in Sect. 2.4.4, the incorporation of a large bandgap buffer layer as in the case of n-ZnS increases the possibility of absorbing high-energy photons at the blue end of the electromagnetic spectrum coupled with the prospect of the incorporation of the impurity PV effect and impact ionisation aiding photon absorption in the infrared (IR) region. The depletion region in this glass/FTO/n-ZnS/n-CdS/n-CdTe/Au structure may span across the whole device thickness when adequately optimised as

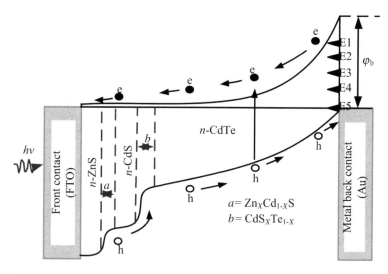

Fig. 7.8 Typical band diagram of glass/FTO/*n*-ZnS/*n*-CdS/*n*-CdTe/Au device configuration

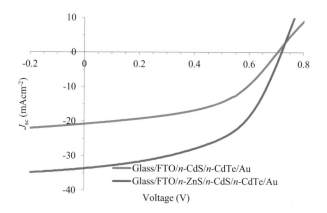

Fig. 7.9 *I-V* characteristics of both glass/FTO/*n*-CdS/*n*-CdTe/Au and glass/FTO/*n*-ZnS/*n*-CdS/*n*-CdTe/Au structures under AM1.5 illumination condition

depicted by the band bending (instigating potential difference and integrated electric field for e-h separation) shown in Fig. 7.8.

The *I-V* characteristics of both glass/FTO/*n*-CdS/*n*-CdTe/Au and glass/FTO/*n*-ZnS/*n*-CdS/*n*-CdTe/Au structure under AM1.5 illumination condition are shown in Fig. 7.9, while Table 7.5 shows the summary of device parameters obtained from *I-V* (both under illuminated and dark conditions) and *C-V* (dark condition). As observed in Table 7.5, it is not surprising that higher J_{sc} above the Shockley-Queisser limit of a single *p-n* junction was observed for glass/FTO/*n*-ZnS/*n*-CdS/*n*-CdTe/Au as compared to the glass/FTO/*n*-CdS/*n*-CdTe/Au configuration. Further to this is the improvement of the V_{oc}, FF and consequently the conversion efficiency

Table 7.5 Summary of device parameters obtained from *I-V* (both under illuminated and dark conditions) and *C-V* (dark condition) for glass/FTO/*n*-CdS/*n*-CdTe/Au and glass/FTO/*n*-ZnS/ *n*-CdS/*n*-CdTe/Au solar cells

CdTe growth voltage (mV)	Glass/FTO/*n*-CdS/*n*-CdTe/Au (two-layer device)	Glass/FTO/*n*-ZnS/*n*-CdS/*n*-CdTe/Au (three-layer device)
I-V under dark condition		
R_{sh} (Ω)	$>10^5$	$>10^5$
R_s (kΩ)	0.50	0.47
log (RF)	3.9	4.8
I_o (A)	1.0×10^{-9}	1.0×10^{-9}
n	1.86	1.60
ϕ_b (eV)	>0.81	>0.82
I-V under AM1.5 illumination condition		
I_{sc} (mA)	0.65	1.07
J_{sc} (mAcm^{-2})	20.70	34.08
V_{oc} (V)	0.72	0.73
Fill factor	0.50	0.57
Efficiency (%)	7.50	14.18
C-V under dark condition		
$\sigma \times 10^{-4}$ (Ω.cm)$^{-1}$	2.85	8.82
$N_D - N_A$ (cm^{-3})	3.10×10^{14}	7.79×10^{14}
μ (cm^{2}V^{-1} s^{-1})	5.74	7.07
C_o (pF)	330	280
W (nm)	926.7	1092.2

(see Table 7.5). Likewise, both the mobility and the depletion width are comparatively higher owing to improved electronic properties.

7.6   Characterisation of Glass/FTO/*n*-CdS/*n*-CdTe/*p*-CdTe/ Au Solar Cell

The motivation for the incorporation of thin *p*-CdTe layer to the glass/FTO/*n*-CdS/ *n*-CdTe/Au configuration as discussed in Sect. 7.4 is for the formation of a depletion region and band bending at the *n*-CdTe/*p*-CdTe interface with very low lattice mismatch. Further to this, the minimisation of contact resistance [26, 27], enhancement of band bending by pinning the Fermi level (FL) close to the valence band [7] and improvement of the reproducibility of the devices were all expected. This configuration is expected to increase photo-generated charge carriers by minimising recombination within the bulk of the configuration and enhance J_{sc}. It should be noted that pinning the FL close to the valence band of semiconductor materials such as CdTe can be achieved through surface etching as demonstrated in the literature [28, 29], incorporation of p$^+$ dopant and other surface treatments [30].

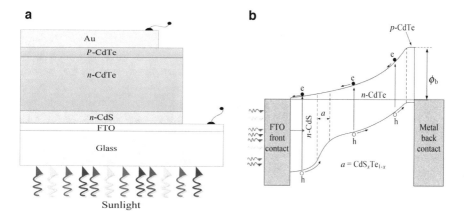

Fig. 7.10 (**a**) Schematic diagram and (**b**) energy band diagram of the glass/FTO/n-CdS/n-CdTe/p-CdTe/Au thin-film solar cell

Figure 7.10a shows the schematic diagram of the preliminary device structure fabricated and tested in this work. The full device consisted of three semiconducting layers as in glass/FTO/n-CdS ($\sim$150 nm)/n-CdTe ($\sim$1144 nm)/p-CdTe ($\sim$35 nm)/Au. The thicknesses used for different layers are also indicated, and a thin n-CdTe layer was primarily used due to our previous experimental observations reported in Sect. 7.4. This work indicated that superior solar cells arise from device configurations with n-type CdTe rather than p-type CdTe as the bulk of the layer. The energy band diagram of this device structure is shown in Fig. 7.10b, and a thin p-CdTe ($\sim$35 nm) layer was used only to fix the Fermi level closer to the valence band maximum so that a healthy band bending occurs throughout the device structure. It should be noted that the n- and p-CdTe layers were grown very close to V_i using the same bath in order to keep the crystallinity high, to remove additional interface states and therefore to minimise native defects. After growing n-type CdTe at a $V_g > V_i$, the growth voltage was simply changed to $V_g < V_i$, to grow a thin p-type CdTe layer. This doping is simply achieved by changing the stoichiometry of the layers rather than using any external doping agent.

7.6.1   Results and Discussion

In a typical experimental sample, about 36, 2 mm diameter, Au contacts can be made. All these devices were measured using a commercially available fully auto-mated I-V system. Solar cell measurements were carried out under AM1.5 illumi-nation, and the system was calibrated using a standard reference cell RR267MON.

The statistics of efficiency values are plotted in Fig. 7.11, and the scatter of efficiency is wide for this preliminary work. While the work is progressing to improve reproducibility, consistency and uniformity, the device parameters of the

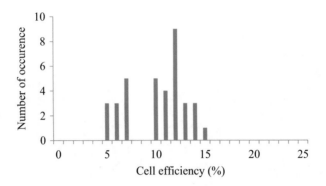

Fig. 7.11 Statistics showing the cell efficiency distribution in this preliminary work

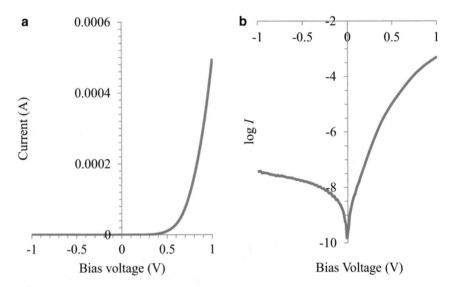

Fig. 7.12 Current-voltage curves plotted using (**a**) linear-linear and (**b**) log-linear scales for the best solar cell measured in the batch

best device observed with the efficiency of 15.3% are studied in detail and presented in this section.

Figure 7.12 shows the *I-V* characteristics of the best device measured under dark condition. These were plotted in both linear-linear and log-linear scales, in order to extract essential device properties, and these parameters are summarised in Table 7.6. Rectification factor (RF) at 1.00 V exceeds four orders of magnitude, indicating the excellent rectifying quality of this device. Potential barrier height available in this structure for electron transport is greater than 0.80 eV, but the real value is underestimated due to a large ideality factor of 1.86.

To further investigate the properties of the depletion region and the doping concentration of the material, dark *C-V* measurements were carried out at a signal

Table 7.6 Summary of the device parameters measured under dark condition for the highest-efficiency solar cell

RF	n	I_0 (A)	ϕ_b (eV)	R_s (kΩ)	R_{sh} (Ω)
$10^{4.1}$	1.86	10^{-9}	>0.80	0.50	7.2×10^5

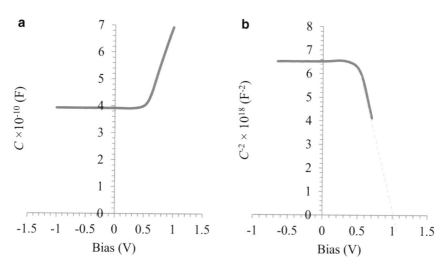

Fig. 7.13 (a) Capacitance-voltage plot and (b) Mott-Schottky plot for the cell with highest conversion efficiency of 15.3%

Table 7.7 Summary of materials and device parameters obtained for 15.3% efficiency solar cell

Zero-bias capacitance (pF)	Geometrical capacitance (pF)	Built-in potential (V_{bi}) eV	Doping density ($N_D - N_A$) cm^{-3}
395	238	~1.00	6.67×10^{14}

frequency of 1.0 MHz. Figure 7.13 shows the variation of capacitance as a function of bias voltage and corresponding Mott-Schottky plot for the highest performing device. It is clear from the shape of the C-V curve that the device is fully depleted at reverse-biased and close to zero-biased voltages. As the device is forward biased and voltage is increased, the depletion width, W, becomes equal to the thickness of the device which is ~1.2 μm, around a forward-bias voltage of ~0.5 V. After this voltage, the capacitance increases with the increasing forward-bias voltage, gradually reducing the depletion region. This variation behaves according to Mott-Schottky theory and provides estimates for V_{bi} and ($N_D - N_A$) for the n-CdTe layer. The data estimated are given in Table 7.7, and the V_{bi} of ~1.0 eV and ($N_D - N_A$) ~6.7 × 10^{14} cm^{-3} are most acceptable for this device. V_{bi} ~1.0 eV corresponds to a potential barrier height of ~1.10 eV, and the doping concentration in the region corresponds to the high-efficiency CdTe devices (~1.0 × 10^{14}–5 × 10^{15} cm^{-3}) [26, 31, 32].

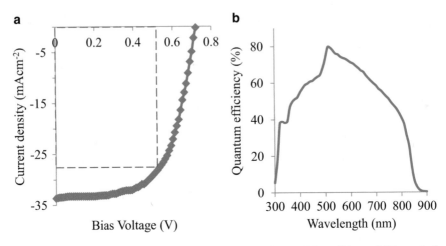

Fig. 7.14 (**a**) *I-V* plot of the cell with highest efficiency under AM1.5 condition and (**b**) a typical IPCE curve for cells with efficiency of ~10%

Figure 7.14a shows the *I-V* curve recorded under the AM1.5 condition for the best solar cell measured, giving V_{oc} = 730 mV, J_{sc} = 33.8 mAcm^{-2}, FF = 0.62 and conversion efficiency of 15.3%. Figure 7.14b shows an incident photon-to-current efficiency (IPCE) curve for a device with efficiency ~10% from this batch. This shows the PV-active nature of these devices in the wavelength range between 300 and 880 nm, with a peak ~500 nm.

As indicated by the statistics shown in Fig. 7.11, for these preliminary devices, it is clear that the uniformity of device parameters and consistency need improving. However, the best devices showing 15.3% efficiency for this three-layered *n-n-p* device show the high potential of achieving further improvements in materials and processing optimisation. The series resistance of 500 Ω measured for the best device is high, and reduction of this should further improve the FF and the short-circuit current density.

This work is only the first step towards the development of graded bandgap solar cells using only three layers. Bandgap grading takes place at only in the CdS/CdTe interface during CdCl$_2$ treatment. In view of increasing conversion efficiency and reducing the R_s, copper-gold contacts were evaporated on a direct replica of the glass/FTO/*n*-CdS/*n*-CdTe/*p*-CdTe to achieve glass/FTO/*n*-CdS/*n*-CdTe/*p*-CdTe/ Cu-Au devices. A champion efficiency of 18.5% was observed under AM1.5 conditions (see Fig. 7.15 and Table 7.8). But due to instability and reproducibility issues, the results have not been published yet until the electronic parameters of the fabricated cells can be stabilised and further experimentation performed.

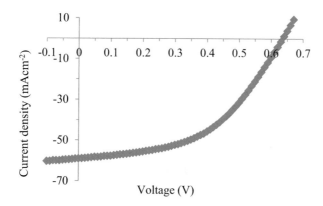

Fig. 7.15 Typical *I-V* curve for the 18.5% conversion efficiency observed under AM1.5 illuminated condition

Table 7.8 Typical cell electronic property for the 18.5% conversion efficiency observed under both AM1.5 and dark conditions

AM1.5 condition		*I-V* dark condition		*C-V* dark condition	
V_{oc} (V)	0.64	R_{sh} (Ω)	10^6	C_0 (pF)	160
J_{sc} (mAcm^{-2})	58.9	R_s (Ω)	925	V_{bi} (V)	>0.65
FF	0.50	RF	3.5	slope C^{-2}	7.14×10^{19}
η (%)	18.5	I_o or I_s	3.16×10^{-9}	N_D (cm^{-3})	1.82×10^{14}
		n	1.68		
		ϕ_b (eV)	>0.80		

7.6.2   Summations

The work presented here successfully combined the knowledge acquired from two different research fronts: electrodeposition of semiconductors and the development of graded bandgap device structures. Without making any ambiguous assumptions, *n-n-p* device structures were fabricated using well-studied electroplated *n*-CdS, *n*-CdTe and *p*-CdTe layers. As expected, this preliminary study produced all PV-active devices showing the best efficiency of 15.3%. *I-V*, *C-V* and IPCE measurements confirm promising devices capable of developing into multilayer graded bandgap solar cells to harvest most of the photons available to achieve highest possible conversion efficiency.

7.7 Effect of Fluorine Doping of CdTe Layer Incorporated in Glass/FTO/*n*-CdS/*n*-CdTe/*p*-CdTe/Au

After the structural, optical, morphological, compositional and electrical study on the CdTe layers whose baths were doped with different concentrations of fluorine was completed (see Sect. 6.6.1), similar CdTe layers of ~1500-nm-thick were grown from 0 to 50 ppm F-doped baths on pretreated glass/FTO/CdS substrate and capped with electrodeposited *p*-type CdTe from the 0 ppm F-doped bath.

The 120-nm-thick CdS layer utilised in the glass/FTO/*n*-CdS/*n*-CdTe/*p*-CdTe configuration was electrodeposited from an electrolytic bath containing 0.03 M ammonium thiosulphate ((NH_4)$_2$$S_2$$O_3$) and 0.3 M cadmium chloride hydrate ($CdCl_2$·xH_2O) at an optimised cathodic voltage of 1200 mV based on morphological, compositional, structural, electronic and optical analysis [33]. The glass/FTO/CdS layer was $CdCl_2$ treated at 400 °C in air prior to the deposition of CdTe layers. The incorporation of the *p*-CdTe layer is to force the Fermi level very close to the valence band. The schematic diagram and the band diagram of the fabricated solar cells are shown in Fig. 7.16.

It is important to note that during the $CdCl_2$ treatment of *n*-CdS/*n*-CdTe/*p*-CdTe structure, there is an interdiffusion of Te and S which results in the formation of CdS_xTe_{1-x} intermediate material at the CdS/CdTe interface [34, 35]. This intermediate material is expected to have a bandgap between that of CdS and CdTe, thus causing a grading in bandgap between CdS and CdTe. The small bowing effects of the bandgap as reported in the literature [36] do not have considerable effect on the shape of the large band bending.

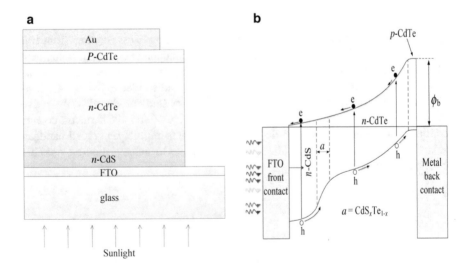

Fig. 7.16 (**a**) Schematic diagram and (**b**) the band diagram of the glass/FTO/*n*-CdS/*n*-CdTe/*p*-CdTe/Au thin-film solar cell

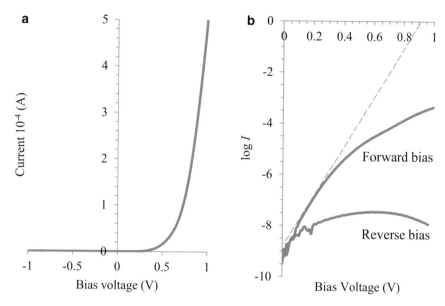

Fig. 7.17 (a) Typical linear-linear *I-V* curve and (**b**) semilogarithmic current vs. voltage curve measured under dark condition for glass/FTO/*n*-CdS/*n*-CdTe/*p*-CdTe/Au devices (CdTe was grown from 20 ppm doped CdTe bath)

The structure (glass/FTO/*n*-CdS/*n*-CdTe/*p*-CdTe) was heat treated with CdCl$_2$ at 400 °C for 20 min in air. The surface was etched using a solution containing K$_2$Cr$_2$O$_7$ and concentrated H$_2$SO$_4$ for acid etching and a solution containing NaOH and Na$_2$S$_2$O$_3$ for basic etching. 2 mm diameter Au contacts were evaporated at a vacuum pressure of ~10^{-4} Nm^{-2}. The glass/FTO/*n*-CdS/*n*-CdTe/*p*-CdTe/Au devices were analysed using both *I-V* and *C-V* characteristic measurements to determine their device parameters. Typical linear-linear and log-linear *I-V* curves measured under dark condition for a device incorporating CdTe from the 20 ppm F-doped CdTe bath are shown in Fig. 7.17.

The linear *I-V* curve under AM1.5 illuminated condition is shown in Fig. 7.18, the typical Mott-Schottky plot of glass/FTO/*n*-CdS/*n*-CdTe/*p*-CdTe/Au layer is shown in Fig. 7.19, while the device parameters are presented in Table 7.9 for comparison. The effective Richardson constant (A^*) has been calculated to be 12 Acm^{-2} K^{-2} for CdTe.

As observed in the dark *I-V* section of Table 7.9, the shunt resistance (R_{sh}) was comparatively high across all F-doped CdTe devices but low for the 50 ppm F-doped materials. Low R_{sh} values of a solar device as explained by Soga (2004) can be directly related to low semiconductor material quality which might be due to the inclusion of voids, gaps and high dislocation density within the semiconductor material [19]. It should be noted that R_{sh} is more dominant in low light conditions and may result in the reduction in fill factor (FF) and the open-circuit voltage (V_{oc}) [37]. It is interesting to know that this observation correlates with the optical and

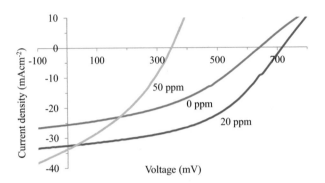

Fig. 7.18 Linear *I-V* curves under AM1.5 for glass/FTO/*n*-CdS/*n*-CdTe/*p*-CdTe/Au solar cells fabricated with CdTe from 0, 20 and 50 ppm fluorine-doped CdTe baths

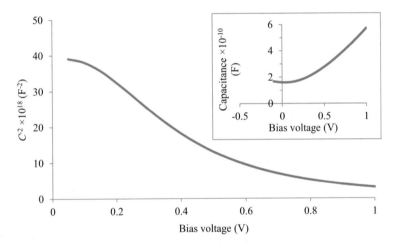

Fig. 7.19 A typical Mott-Schottky plot of glass/FTO/*n*-CdS/*n*-CdTe/*p*-CdTe/Au structure (the *n*-CdTe grown from 20 ppm F-doped CdTe bath). Inset shows the variation of capacitance as a function of bias voltage

morphological summaries on the CdTe material property presented in Sects. 3.2 and 3.3, respectively.

The CdTe layers grown with 0, 5, 10 and 20 ppm F-doped baths show good rectification factor (RF) with values higher than three orders of magnitude, while a drastic reduction in the RF to $\sim 10^{0.9}$ is recorded for device samples with a higher F-doping concentration above 20 ppm. As reported by Dharmadasa 2013, the minimum of $\sim 10^3$ RF value is sufficient for an efficient solar cell [15]. However, higher RF values show high-quality rectifying property of the devices and are desirable for solar cells with very high efficiencies. The drastic reduction in the RF value can be attributed to the deterioration of the CdTe material at higher F-doping concentrations.

Table 7.9 Device parameters from I-V (dark condition), I-V (illuminated at AM1.5) and C-V (dark condition) measurements

F-doping concentration in CdTe bath (ppm)	0	5	10	20	30	50
I-V under dark condition						
R_{sh} (Ω)	$\sim 10^6$	$\sim 10^6$	$\sim 10^6$	$\sim 10^6$	$\sim 10^5$	$\sim 10^3$
R_s (kΩ)	0.52	2.07	5.33	0.50	2.63	0.22
log (RF)	4.8	4.4	3.2	4.2	2.2	0.9
I_o (A)	5.01×10^{-10}	3.16×10^{-10}	3.98×10^{-10}	1.58×10^{-9}	1.58×10^{-7}	7.94×10^{-6}
n	1.60	1.70	1.70	1.60	>2.00	>2.00
ϕ_b (eV)	>0.82	>0.83	>0.83	>0.80	>0.67	>0.57
I-V under AM1.5 illumination condition						
I_{sc} (mA)	0.80	0.89	0.92	1.06	1.06	1.06
J_{sc} (mAcm^{-2})	25.5	28.3	29.3	33.8	33.8	34.4
V_{oc} (V)	0.64	0.62	0.67	0.73	0.62	0.35
Fill factor	0.43	0.43	0.45	0.50	0.44	0.37
Efficiency (%)	7.01	7.56	8.83	12.32	9.21	4.37
DC conductivity and *C-V* under dark condition						
$\sigma \times 10^{-4}$ (Ω.cm)$^{-1}$	1.03	1.12	1.73	1.75	2.23	2.31
$N_D - N_A$ (cm^{-3})	9.4×10^{14}	1.7×10^{14}	1.9×10^{14}	1.8×10^{14}	5.2×10^{15}	7.7×10^{16}
μ (cm^{2}V^{-1}s^{-1})	0.69	4.23	5.61	5.94	0.27	0.02

Furthermore, the device samples grown from baths containing 0 to 20 ppm F-doping concentration show an ideality factor (n) between 1.00 and 2.00, while higher n values are observed at higher F-doping concentration. This indicates that the current transport mechanism of devices grown from electrodeposition baths containing between 0 and 20 ppm F-doping concentration is dominated by both thermionic emission and recombination and generation (R&G) processes in parallel [38]. The ideality factor value, >2.00, observed for devices grown from CdTe baths with F-doping concentration above 20 ppm, shows that R&G process dominates the current transport mechanisms and, in turn, causes a reduction in the estimated barrier height ϕ_b as observed in Table 7.9.

As observed in Fig. 7.18 and the linear I-V curves (under AM1.5) section of Table 7.9, high short-circuit current density (J_{sc}) was observed with increasing F-doping in all electrolytic baths. The observed J_{sc} value was higher than the Shockley-Queisser limit for a single p-n junction [11] due to the multilayer and multi-junction n-n-p [12] device configuration. Further increase in the F-doping concentration above 20 ppm in the electrolytic bath shows increasing J_{sc} but also a reduction in the V_{oc}, FF and the overall conversion efficiency (η).

Figure 7.19 shows a typical Mott-Schottky plot of glass/FTO/n-CdS/n-CdTe/p-CdTe/Au structures with CdTe layers grown using the 20 ppm F-doped CdTe bath for the growth of n-CdTe layer. The C-V measurements were performed in dark condition at a bias range of -1.00 to 1.00 V with 1 MHz AC signal at 300 K. The built-in potential (V_{bi}) and doping density (N_D) in this configuration were determined using the Mott-Schottky plot as shown in Fig. 7.19 and described in Sect. 3.5.2. The carrier mobility $\mu_\perp$ was calculated with the assumption that all donor atoms (N_D) are ionised at room temperature, therefore, $n \approx N_D$ (see Equation 3.57). All the calculated parameters are shown in Table 7.9.

As observed in Table 7.9, the N_D for devices grown using the 0–20 ppm F-doping concentration was $\sim 10^{14}$ cm^{-3}. An increase in the N_D to $\sim 10^{17}$ cm^{-3} was observed above 20 ppm F-doping concentration. Investigation on the optimum N_D of CdTe has been reported in the literature to be $\sim 10^{14}$ [15, 18, 39]. An increase in the mobility was observed within the range of 5–20 ppm F-doping concentration. Above this F-doping concentration range, a reduction in the mobility was observed. The reduction in mobility might be due to the presence of high concentration of defects (R&G centres) as depicted by the high ideality factor on the devices.

7.7.1 Summations

In this work, we have explored the effect of fluorine doping (in CdTe electrolytic bath) on CdTe layer as it affects its structural, optical, morphological and compositional properties. Further investigation on the electronic property of the CdTe layer was also carried out by incorporating the layer into a glass/FTO/n-CdS/n-CdTe/p-CdTe/Au device configuration. An optimal F-doping concentration of ~ 20 ppm was observed under all material and device characterisations with a DC conductivity of 1.75×10^{-4} $(\Omega.\text{cm})^{-1}$, short-circuit current density J_{sc} of 33.76 (mAcm^{-2}), doping density N_D of 1.8×10^{14}, mobility $\mu_\perp$ of 5.94 $\text{cm}^2\text{V}^{-1}\text{s}^{-1}$ and conversion efficiency of 12.3%. These observed parameters can still be improved through more precise processing steps, improved material quality and improved metal/semiconductor contact property amongst others. Development of multilayer graded bandgap devices is in progress towards achieving the highest possible efficiency.

7.8 Summary of the Effects of Fluorine, Chlorine, Iodine and Gallium Doping of CdTe

Further to the impact of F, Cl, I and Ga doping of CdTe on its material property as discussed in Sect. 6.6, ~1500-nm-thick doped CdTe layers were incorporated in glass/FTO/*n*-CdS/*n*-CdTe/*p*-CdTe/Au. It should be noted that the optimal CdTe under all the explored doping was n-type, while the thicknesses of the *n*-CdS and *p*-CdTe were ~120 nm and ~30 nm, respectively, for this configuration. The processing steps and the measured electronic parameters for different dopings Cl, I and Ga were not presented in this book to avoid tautology following the explicit documentation for CdTe/F as presented in Sect. 7.6. Table 7.10 shows the summary of device parameters from *I-V* under dark and AM1.5 illuminated conditions and *C-V* (dark condition) measurements incorporating F, Cl, I and Ga optimal doping in *n*-CdTe.

It should be noted that the Cl doping of CdTe was not in ppm level but rather as the precursor of the electrolyte from which CdTe was deposited. The optimum cathodic voltage is 1360 mV. As shown in Table 7.10, the electronic properties of the fabricated devices are relatively similar except for a few pointers since they all satisfy requirements of high-efficiency solar cells such as *RF* of 10^3 and above and *n* value <2.00 which signifies the domination of the current transport mechanism by both thermionic emission and R&G. Relatively, current density J_{sc} across all the explored devices shows values higher than the value for single *p-n* junction as suggested by Shockley-Queisser [11] due to the multilayer configuration of the explored devices [12]. Under AM1.5 illuminated condition, the lowest efficiency was observed with devices incorporating I-doped n-CdTe layer, while the highest conversion efficiency was recorded for the device incorporating Ga-doped n-CdTe with comparatively higher J_{sc}, FF and V_{oc}. The high efficiency in the devices incorporating Ga-doped *n*-CdTe might be due to the improvement in both the material and electronic properties due to possible reduction or elimination of Te precipitation during growth which is a possible reason for the high mobility as shown in Table 7.10. It should also be noted that the comparatively low doping observed for the device incorporating Cl-doped *n*-CdTe devices might be because the *n*-CdTe layers were grown in proximity to stoichiometry as depicted in Sect. 6.6.2. It is, therefore, safe to say that although there has been no clear evidence of the better devices, it could still be pointed out that Ga-doped CdTe shows the most promising electronic parameters.

Table 7.10 Summary of device parameters obtained from *I-V* curves measured under dark and AM1.5 illuminated condition and *C-V* (dark condition) measurements incorporating F, Cl, I and Ga optimal doping of CdTe as *n*-CdTe in glass/FTO/*n*-CdS/*n*-CdTe/*p*-CdTe/Au configuration

Dopant	F	Cl	I	Ga
Opti. doping conc. (ppm)	20	1.5 M CdCl$_2$ (at 1360 mV)	5	20
I-V under dark condition				
R_{sh} (Ω)	>10^6	>10^6	>10^6	>10^6
R_s (kΩ)	0.51	0.46	0.90	0.52
log (RF)	4.2	4.6	3.3	4.4
I_o (A)	1.58 × 10^{-9}	1.02 × 10^{-9}	3.16 × 10^{-9}	5.01 × 10^{-10}
n	1.60	1.60	1.80	1.61
ϕ_b (eV)	>0.80	>0.81	>0.73	>0.82
I-V under AM1.5 illumination condition				
I_{sc} (mA)	1.06	0.96	0.94	1.09
J_{sc} (mAcm^{-2})	33.76	30.57	29.94	34.71
V_{oc} (V)	0.73	0.71	0.66	0.73
Fill factor	0.50	0.56	0.50	0.57
Efficiency (%)	12.32	12.16	9.88	14.44
C-V under dark condition				
σ × 10^{-4} (Ω.cm)$^{-1}$	1.75	1.09	2.43	1.22
N_D (cm^{-3})	1.80 × 10^{14}	3.90 × 10^{13}	6.69 × 10^{14}	2.01 × 10^{14}
μ (cm^{2}V^{-1} s^{-1})	6.07	17.45	2.27	37.83
C_o (pF)	160	280	400	170
W (nm)	1911.4	1092.2	774.2	1798.9

7.9   Effect of Cadmium Chloride Post-growth Treatment pH

Although efficiency stagnation in the cadmium sulphide-/cadmium telluride (CdS/CdTe)-based solar cell has been reported in the literature for the past 20 years prior to the recent improvement in both material and processing issues, post-growth treatment (PGT) has been documented as one of the most crucial processing step towards enhancing solar to electrical energy conversion efficiency. With properties such as grain growth, recrystallisation, improved stoichiometry, grain boundary passivation and optimisation of doping concentration [1], amongst other advantages attributed to the PGT of CdS/CdTe, PGT has been the focus of many researchers. Research focus has been turned towards identifying the best chlorine-based gas or salt solution either in aqueous or methanol in which the highest efficiency can be achieved [5, 40–43], the best application method [44] and also the optimisation of both annealing temperature and time [45]. So far, the effects of the pH values of the chlorine salt solution have been often overlooked. With an emphasis on cadmium chloride ($CdCl_2$) PGT, these sets of experiments focus on the effect of PGT solution treatment pH on both the material and device properties of CdS-/CdTe-based solar cells.

7.9.1 Fabrication and Treatment of Glass/FTO/n-CdS/ n-CdTe/p-CdTe/Au

The initial preparation of the glass/FTO substrate (with 5×4 cm^2 dimension) was performed as described in Sect. 5.2.2. A 120-nm-thick CdS layer was grown on the glass/FTO strip at 1200 mV cathodic voltage. After CdS deposition, the glass/FTO/CdS was rinsed in DI water and dried in a nitrogen stream, and $CdCl_2$ was applied as described in Sect. 7.2. Within the experimental constrain of this work, stoichiometric CdTe was observed at 1370 mV, while both the n-CdTe and p-CdTe for these sets of experiments were grown at 1375 mV and 1365 mV, respectively. Utilising a continuous deposition process, 1200-nm-thick n-CdTe followed by a 30-nm-thick p-CdTe was grown to achieve glass/FTO/n-CdS/n-CdTe/p-CdTe configuration. The incorporation of the comparatively thin p-CdTe layer was necessitated to force the Fermi level close to the valence band and also to reduce the contact resistance at the metal/semiconductor interface [26]. However, an increase in the p-CdTe thickness in this configuration causes a detrimental effect on the device parameters due to increase in defect density associated with p-CdTe layers [46]. Both the CdS and the CdTe layers were grown using optimised growth voltages explored in Chaps. 4 and 5, respectively.

Post-growth treatment commences immediately after the growth of CdTe resulting into glass/FTO/n-CdS/n-CdTe/p-CdTe configuration. The glass/FTO/n-CdS/n-CdTe/p-CdTe layer is rinsed in DI water to remove loose Cd, Te or CdTe and dried in a stream of nitrogen gas. The 5×4 cm^2 glass/FTO/n-CdS/n-CdTe/p-CdTe was cut into five strips (of the 1×4 cm^2 area) from the glass side, rinsed thoroughly in running DI water to wash off the glass and dried in a stream of nitrogen gas. Prior to the application of $CdCl_2$ treatment, 0.1 M $CdCl_2$ was dissolved in 80 mL of DI water in a 100 mL glass beaker at room temperature. To achieve homogeneity, the solution was stirred for 60 min, and 20 mL of the solution was poured into four different 25 mL glass beakers. The beakers were labelled A to D with the solution contained in beaker A being the most acidic with a pH of 1.00 ± 0.02, beaker B with pH 2.00 ± 0.02 and beaker C with pH 3.00 ± 0.02, and beaker D was left as-prepared with pH of ~4.02 ± 0.02. It should be noted that the acidity level of the $CdCl_2$ solution contained in the 25 mL beaker was adjusted using dilute HCl.

The $CdCl_2$ solution with different pH treatments was applied by adding few drops on each strip labelled A to D at this point based on the pH of the solution in which they were treated, while the fifth strip E was left as-deposited. Each strip was allowed to air-dry before heat treating at 420 °C for 20 min in air except the as-deposited strip E. Afterwards, each strip of glass/FTO/n-CdS/n-CdTe/p-CdTe layers was rinsed in running DI water, dried in stream of nitrogen gas and etched in accordance with the description made in Sect. 7.2. 100-nm-thick gold (Au) contacts were evaporated on the glass/FTO/n-CdS/n-CdTe/p-CdTe using a 3-mm-diameter mask. The fabricated devices were analysed using both current-voltage and capacitance-voltage charac-teristic measurements to determine their device parameters. (It should be noted that the CdCl$_2$ post-growth treatment referred to in this section denotes CdCl$_2$ treatment and heat treatment at 420 °C for 20 min in the air.)

7.9.2 Effect of CdCl$_2$ Treatment pH on the Material Properties of Glass/FTO/n-CdS/n-CdTe/p-CdTe Layers

7.9.2.1 Optical Property Analysis

Further to the experimental details as discussed in Sect. 3.4.6, Fig. 7.20a shows the plot of A^2 against photon energy $h\nu$. Figure 7.20b shows the graph of absorption edge slope against the CdCl$_2$ post-growth treatment pH. The optical energy bandgap of the as-deposited and the CdCl$_2$-treated CdTe layers were obtained by extrapolat-ing the linear portion of the curve to $A^2 = 0$. From observation, it could be said that the optical bandgap lies within the 1.45 ± 0.01 eV for the as-deposited and all the CdCl$_2$-treated CdTe layers. This observed bandgap shows comparability with the standard bulk CdTe bandgap of 1.45 eV. More importantly, the absorption edge

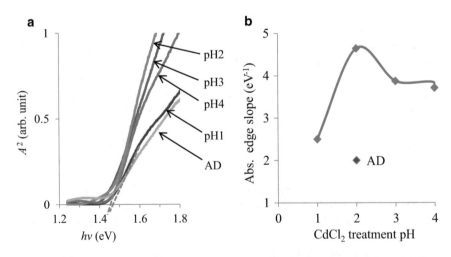

Fig. 7.20 (a) Optical absorption spectra for electrodeposited glass/FTO/n-CdS/n-CdTe thin films treated with different CdCl$_2$ at different pH values, (b) absorption edge slope against PGT CdCl$_2$ pH CdTe thin films

slope as shown in Fig. 7.20b can be related to semiconductor layer quality as discussed in the literature [5, 47]. As expected, an improvement in the absorption edge of the as-deposited glass/FTO/n-CdS/n-CdTe was observed after CdCl$_2$ treatment at different pHs. This improvement has been well documented in the literature [1, 41, 44].

The steepest absorption edge slope was observed at pH2, while the lowest absorption edge slope was observed at pH1. This reduction in the absorption edge slope might be due to the reduction in the quality of the glass/FTO/n-CdS/n-CdTe layer as a result of the harshness of the acidic CdCl$_2$ treatment by possible dissolution of Cd from CdTe at high acidity. It should be noted that only the bandgap of CdTe was observable rather than that of the incorporated n-CdS layer nor the bowing effect of CdSTe alloy [36] due to the thickness of the CdTe layer.

7.9.2.2 Morphological and Compositional Analysis

Figure 7.21a–d shows the SEM micrographs of glass/FTO/n-CdS/n-CdTe in the as-deposited CdCl$_2$ treated at pH1, pH2 and pH4, respectively, while Fig. 7.21e, f shows the energy-dispersive X-ray (EDX) spectra of point identification on the glass/FTO/n-CdS/n-CdTe treated with pH1 CdCl$_2$ solution. The layer treated with pH3 CdCl$_2$ was excluded due to its high comparability of morphological properties with pH4. The as-deposited glass/FTO/n-CdS/n-CdTe layer as depicted in Fig. 7.21a shows cauliflower-type morphology which is formed by the agglomerations of small grains. Most importantly, full coverage of the underlying glass/FTO/n-CdS layers was observed.

After CdTe treatment at all the explored pH in this work, an increase in grain growth was observed, which is in accord with the literature. The layers treated with pH2 CdCl$_2$ showed a slightly bigger grain size as compared to the layers treated with pH4 CdCl$_2$ as shown in Fig. 7.21. The glass/FTO/n-CdS/n-CdTe layers treated with pH1 CdCl$_2$ as illustrated in Fig. 7.21b show deterioration of the glass/FTO/n-CdS/n-CdTe layer with the presence of pinholes and the accumulation of non-uniform strands on the grains.

With further investigation on the composition of the strands using EDX as shown in Fig. 7.21e, f, it was observed that the strands show an atomic composition of 75.6% for Te and 24.4% for Cd. The presence of the Te-rich strands has also been documented in the literature [48, 49], and it is well known that an introduction of an acidic media to CdTe attacks Cd preferentially leaving rich Te surface [7, 50]. The unmarked EDX peaks in Fig. 7.21e, f at ~2.5, 3.5 and 4.5 keV are for S, Cd and Te, respectively [51, 52]. This observation signifies the detrimental effect of pH1 CdCl$_2$ for post-growth treatment on the material quality of the glass/FTO/n-CdS/n-CdTe all-electrodeposited layers and may result in the reduction in the device quality.

Figure 7.22 shows the graph of Cd/Te atomic composition against the acidity of the CdCl$_2$ PGT of glass/FTO/n-CdS/n-CdTe on a 6 × 6 μm^2 area obtained using EDX. As shown in Fig. 7.22, reduction in the atomic concentration of Cd with respect to AD material was observed after CdCl$_2$ treatment at all the pH explored.

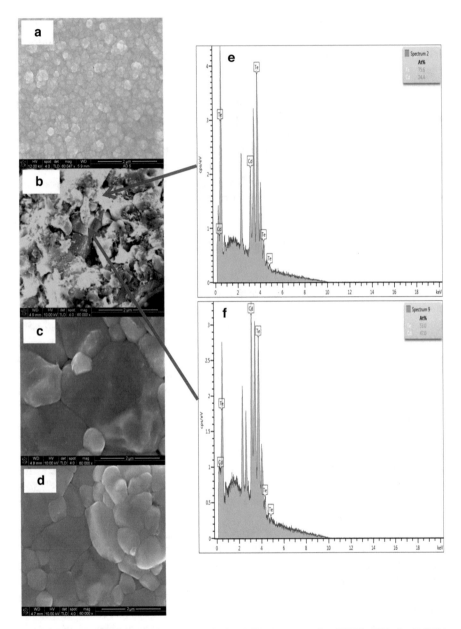

Fig. 7.21 (a) SEM micrograph of as-deposited n-CdTe grown on glass/FTO/n-CdS, (**b–d**) SEM micrographs for glass/FTO/n-CdS/n-CdTe layers treated with PGT treated with CdCl$_2$ at pH1, pH2 and pH4, respectively, while (**e, f**) are the EDX point micrograph on layers treated at pH1

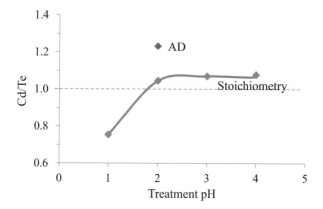

Fig. 7.22 Graphical representation of percentage atomic composition ratio of Cd to Te atoms for CdCl$_2$-treated CdTe layer after different PGT pH as obtained from EDX micrographs

The reduction in the Cd atomic concentration and shift towards 1:1 ratio of Cd to Te can be attributed to CdCl$_2$ treatment at favourable pH [1]. At pH1, an increase in the Te atomic concentration was observed due to harsh effect of highly acidic CdCl$_2$ on elemental Cd. This observation can be related to the Te richness obtained after wet acid etching of CdTe layer as reported in the literature [7].

7.9.2.3   Structural Analysis

The analysis was aimed at identifying the effect of CdCl$_2$ post-growth treatment pH on XRD peak intensity, crystallinity, crystallite size, preferred phase and orientation of the glass/FTO/n-CdS/n-CdTe layers. Figure 7.23a shows the graph of XRD diffraction intensity of the glass/FTO/n-CdS/n-CdTe layers treated at different CdCl$_2$ pH against 2θ angle. Figure 7.23b shows the graph of XRD peak intensity and crystallite size against the CdCl$_2$ post-growth treatment pH. It should be noted that the stacked XRD micrographs as presented in Fig. 7.23a are to aid the comparability of the peak intensity. From observation, XRD peaks associated with CdTe in their cubic phase ((111)C, (220)C, (311)C) were observed at angles $2\theta \approx 23.8$, $2\theta \approx 38.6°$ and $2\theta \approx 45.8°$, respectively, aside the FTO peaks at $2\theta = 25.42$, $2\theta = 33.11$, $2\theta = 36.57$, $2\theta = 55.06$, $2\theta = 61.12$ and $2\theta = 65.06$ at all the CdCl$_2$ pH treatmentss explored.

   As shown in Fig. 7.23a, it is clear that the preferred orientation of CdTe at all the pH explored is along the cubic (111) plane based on the intensity of its diffraction. Furthermore, an increase in diffraction intensity of the (111)C peak was observed with increasing CdCl$_2$ post-growth treatment pH. The highest diffraction intensity is observed at pH2, and the lowest intensity is observed at pH1 as shown in Fig. 7.23b. This detrimental effect can be further related to harsh etching and sublimation of the CdTe surface, the formation of pinholes, voids and the formation of a CdTe layer rich in Cd or Te with competing phases with CdTe. From

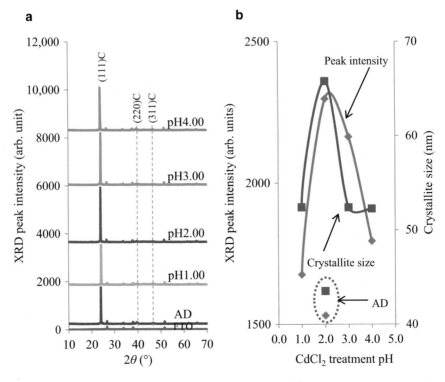

Fig. 7.23 (a) Typical XRD patterns of glass/FTO/n-CdS/n-CdTe layers treated at different $CdCl_2$ pH value, (b) typical plot of CdTe (111) cubic peak intensity and crystallite size against $CdCl_2$ post-growth treatment pH value

Table 7.11 The XRD analysis of glass/FTO/n-CdS/n-CdTe layers treated with $CdCl_2$ at different pH values

Sample	2θ (°)	Lattice spacing (Å)	FWHM (°)	XRD peak intensity	Crystallite size D (nm)	Assignments
AD	23.95	3.716	0.195	1531	43.5	Cubic
pH1.00	24.15	3.681	0.162	1675	52.4	Cubic
pH2.00	23.99	3.707	0.129	2297	65.8	Cubic
pH3.00	24.05	3.696	0.162	2163	52.4	Cubic
pH4.00	23.95	3.714	0.162	1794	52.3	Cubic

observation, no elemental Cd and Te peaks were observable in Fig. 7.23a which might be due to possible overlap with the FTO peaks, although the formation of elemental Te is most likely as suggested by the compositional analysis as will be discussed later in this section. Based on the preferred cubic (111) CdTe peaks, the crystallite size was calculated using Scherrer's equation addressed in Sect. 3.4.2 and Equation 3.8.

Table 7.11 shows the calculated crystallite size and other related properties as obtained from XRD diffraction. From observation, the minimum value of the FWHM and the maximum crystallite sizes were attained at pH2. Away from this pH value, a gradual increase in the FWHM and decrease in the crystallite size were observed. This shows further superior quality of $CdCl_2$ post-growth treatment at pH2 as compared to the other pH values explored. It should be noted that the extracted XRD data from this CdTe work matches the International Centre for Diffraction Data (JCPDS) reference file No. 01-775-2086.

7.9.2.4   Photoelectrochemical (PEC) Cell Study

For this experiment, 1200-nm-thick n-CdTe was grown at 1370 mV on 4×5 cm^2 glass/FTO. This experiment was performed to ascertain the effect of $CdCl_2$ post-growth treatment at different pHs on the conductivity type of the n-CdTe layer utilised in this work. After growth, the glass/FTO/n-CdTe layer was cut into five 1×5 cm^2 pieces and treated with $CdCl_2$ at different pHs prior to heat treatment at 420 °C for 20 min as described in Sect. 7.9.1. Figure 7.24 shows the graph of PEC signal against as-deposited and post-growth-treated n-CdTe at different pH values. As observed from Fig. 7.24, the conductivity type of the as-deposited n-CdTe was retained after $CdCl_2$ treatment at pH2, pH3 and pH4 explored in this work with a slight shift towards the opposing conductivity type except for the layers treated with $CdCl_2$ at pH1. The glass/FTO/n-CdTe layer treated with pH1 shows a conductivity-type transition into p-type. It should be noted that conductivity-type conversion after $CdCl_2$ treatment may be due to doping effect caused by heat treatment temperature, duration of treatment, initial atomic composition of Cd and Te, the concentration of $CdCl_2$ utilised in treatment, defect structure present in the starting CdTe layer and the material's initial conductivity type as documented in the literature [1, 9, 40, 53]. Based on these observations, coupled with the analysis on composition as discussed earlier in this section, it could be said that the compositional alteration of the initial n-CdTe after pH1 $CdCl_2$ treatment might be one of the determining factors in the conductivity-type conversion of the CdTe layers explored in this work.

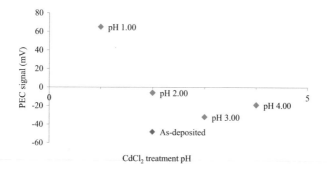

Fig. 7.24   PEC signals for glass/FTO/n-CdTe layers treated with $CdCl_2$ at different pH values

This observation is in accord with the compositional analysis discussed earlier; etching in strong acid (pH1) leads to the preferential removal of Cd, making the surface a Te-rich CdTe layer. As consistently observed, Te-rich CdTe surface exhibits *p*-type electrical conduction.

7.9.2.5   DC Conductivity Study

For this experiment, a 1200-nm-thick n-CdTe layer was grown on glass/FTO, treated with $CdCl_2$ at different pHs after growth and heat treated at 420 °C for 20 min. On the basis of conductivity type as discussed under PEC cell measurement section, gold (Au) contacts were evaporated on the glass/FTO/*p*-CdTe layers, while indium (In) was evaporated on the glass/FTO/*n*-CdTe to form ohmic contacts prior to the *I-V* characterisations of the fabricated cells. From the *I-V* curve generated using Rera Solution PV simulation system, the resistance was calculated as the inverse of the *I-V* slope, and resistivity was calculated as described in Sect. 3.4.8.

Table 7.12 shows the tabulation of properties of the CdTe layers grown on glass/FTO layers, and Fig. 7.25 is an illustration of conductivity and resistance against

Table 7.12 Summary of electrical properties of glass/FTO /*n*-CdTe layers after $CdCl_2$ treatment at different pH values

pH	Resistance R (Ω)	Resistivity $\rho \times 10^4$ ($\Omega \cdot$cm)	Conductivity $\sigma \times 10^{-5}$ $(\Omega \cdot$cm$)^{-1}$
1.00	42.6	1.12	8.97
2.00	9.4	0.25	40.63
3.00	11.3	0.30	33.80
4.00	15.7	0.41	24.33

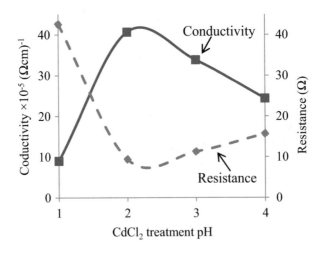

Fig. 7.25   Typical graphs of electrical conductivity and resistance against PGT $CdCl_2$ pH value

CdCl$_2$ treatment pH. It was observed that an increase in the acidity of the CdCl$_2$ post-growth treatment solution increases the conductivity of the CdTe layer with saturation observed at ~pH2.

Increase in the acidity above pH2 shows a reduction in the conductivity which might be due to the p-type conductivity as observed after pH1 CdCl$_2$ treatment as compared to the n-type conductivity as observed after pH2, pH3 and pH4 treatment (see PEC cell measurement result). It is well known that the conductivity and mobility of an n-type semiconductor material are higher than its p-type counterpart [54]. Furthermore, reduction in the conductivity of the CdTe layer might also be due to CdTe layer deterioration as discussed earlier in this section.

7.9.3 The Effect of CdCl$_2$ Treatment pH on Solar Cell Device Parameters

After the optical, morphological, structural and photoelectrochemical properties of the CdTe layers had been analysed, glass/FTO/n-CdS/n-CdTe/p-CdTe/Au devices were fabricated as discussed in Sect. 7.9.1.

7.9.3.1 Current-Voltage Characteristics with Rectifying Contacts

The I-V measurements for the glass/FTO/n-CdS/n-CdTe/p-CdTe/Au devices were performed under both dark and AM1.5 illuminated conditions. Figure 7.26a shows a typical band diagram of the glass/FTO/n-CdS/n-CdTe/p-CdTe/Au thin-film solar cell, while Fig. 7.26b, c show both the linear-linear and log-linear I-V curves of the pH2 CdCl$_2$-treated glass/FTO/n-CdS/n-CdTe/p-CdTe/Au devices, respectively. Figure 7.26d shows the I-V curves taken under AM1.5 illumination condition for the glass/FTO/n-CdS/n-CdTe/p-CdTe/Au devices treated with different pHs of CdCl$_2$, while Table 7.13 shows the summary of the electronic properties of the glass/FTO/n-CdS/n-CdTe/p-CdTe/Au devices fabricated. From the I-V data obtained under dark condition, electrical properties such as the shunt resistance R_{sh}, series resistance R_s, rectification factor RF, reverse saturation current I_o, ideality factor n and the barrier height ϕ_b were derived, while the effective Richardson constant (A^*) for CdTe was calculated using Equation 3.31. As observed in Table 7.13, the R_{sh} was comparatively high for all the pH values explored in this work, but a noticeable reduction of about two orders of magnitude was observed for pH1-treated devices.

It is well known that low R_{sh} can be attributed to the low quality of the semiconductor material [19] which might be due to the inclusion of gaps, voids, pinholes and high dislocation density within the semiconductor layer [19]. Based on this, it can be deduced that the semiconductor material quality at pH1 has been reduced. Interestingly, this observation is in accord with the analytical studies as

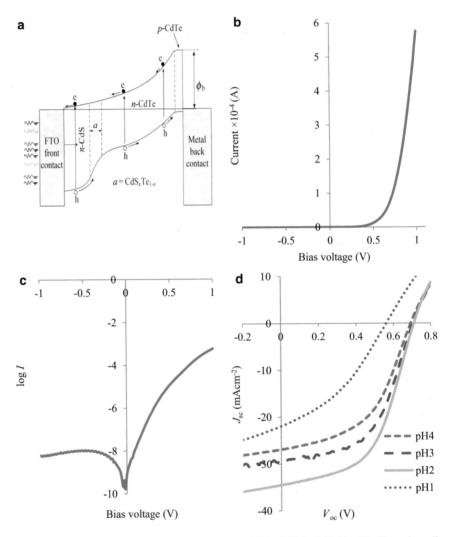

Fig. 7.26 (a) The band diagram of the glass/FTO/n-CdS/n-CdTe/p-CdTe/Au thin-film solar cell. (b) Typical linear-linear I-V curve and (c) semilogarithmic current versus voltage curve measured under dark conditions for glass/FTO/n-CdS/n-CdTe/p-CdTe/Au devices (the layers were treated with pH2 CdCl$_2$). (d) Linear I-V curves of glass/FTO/n-CdS/n-CdTe/p-CdTe/Au under AM1.5 for devices treated with CdCl$_2$ at different pH values

discussed earlier in Sect. 7.9.2. Furthermore, the glass/FTO/n-CdS/n-CdTe/p-CdTe/ Au device activated using CdCl$_2$ at pH2, pH3 and pH4 shows *log RF* values of above 3 with a tendency for achieving highly efficient solar cells [15]. On the contrary, the low log RF values observed for glass/FTO/n-CdS/n-CdTe/p-CdTe/ Au device activated using CdCl$_2$ at pH1 show a lower log RF value of 1.27 which indicates the inability of the fabricated device to achieve high efficiencies.

Table 7.13 Summary of device parameters from I-V characteristics under dark and illuminated (at AM1.5) conditions and C-V measurements under dark conditions

CdCl$_2$ post-growth treatment pH	1.00	2.00	3.00	4.00
I-V under dark condition				
$R_{sh} \times 10^5$ (Ω)	0. 08	10.13	5.72	5.28
$R_s \times 10^3$ (Ω)	1.27	0.47	0.87	0.89
log RF	1.40	4.80	3.50	3.50
$I_o \times 10^{-9}$ (A)	158.49	1.00	3.98	3.16
n	>2.00	1.60	1.86	1.91
ϕ_b (eV)	>0.67	>0.80	>0.77	>0.77
I-V under AM1.5 illuminated condition				
J_{sc} (mAcm^{-2})	21.66	35.03	29.62	27.39
V_{oc} (V)	0.58	0.72	0.71	0.70
FF	0.40	0.52	0.55	0.52
η (%)	5.00	13.10	11.60	10.00
C-V under dark condition				
$\sigma \times 10^{-5}$ (Ω.cm)$^{-1}$	8.97	40.63	33.80	24.33
$N_D \times 10^{14}$ (cm^{-3})	254.00	1.95	3.66	6.67
$\mu_\perp$ (cm^{2}V^{-1}s^{-1})	0.02	13.00	5.76	2.28

This observation might be due to the dominance of the current transport mechanism by recombination and generation (R&G) centres as indicated by an n value >2.00.

It should be noted that for an ideal diode, the ideality factor n is 1.00 indicating the dominance of the current transport mechanism by thermionic emission. But if the ideality factor falls between 1.00 and 2.00, the current transport mechanism consists of both thermionic emission and R&G centres. As reported by Verschraegen et al., the current transport mechanism of a diode with an ideality factor above 2.00 is dominated by high-energy electrons tunnelling through the barrier in addition to both thermionic emission and R&G mechanisms [55]. The current transport mechanism of $n > 2.00$ might result in barrier height ϕ_b reduction as observed in Table 7.13 for the pH1 CdCl$_2$-activated devices.

Under AM1.5 condition as shown in Fig. 7.26d and Table 7.13, the observed J_{sc} of the cell treated using pH1 CdTe post-growth treatment is relatively lower than that of the devices made using pH2, pH3 and pH4. This observation can be related to the high ideality factor as a result of high concentration of R&G centres. It should be noted that the J_{sc} observed in this work is higher than the Shockley-Queisser limit of a single p-n junction [11] due to the multilayer and multi-junction n-n-p device configuration [12]. The explored glass/FTO/n-CdS/n-CdTe/p-CdTe/Au cells were isolated by carefully removing surrounding materials to ensure that there was no peripheral collection as suggested by Godfrey and Green [56]. Using the multilayer configuration, the SHU group has reported 140% IPCE measurement value owing to the incorporation of impurity PV effect and impact ionisation [57], while other independent researchers have also reported EQE values above 100% [58, 59]. Comparatively, similar V_{oc} was observed for glass/FTO/n-CdS/n-CdTe/p-CdTe/Au layers treated with CdCl$_2$ at pH2 to pH4, while a reduction in the V_{oc}, FF and η of the layers treated with pH1 CdTe was observed. These observations were anticipated due to the degradation of the material quality, reduction in crystallinity, conductivity type transition and high resistivity based on the analysis presented in Sect. 7.9.2.

7.9.3.2 Capacitance-Voltage Characteristics of Rectifying Contacts

Figure 7.27a, b show the capacitance-voltage (C-V) and the Mott-Schottky (C^{-2} versus V) plot of the glass/FTO/n-CdS/n-CdTe/p-CdTe/Au device, respectively, with pH2 CdCl$_2$-treated CdTe layers. The properties such as the doping density N_D and mobility μ for glass/FTO/n-CdS/n-CdTe/p-CdTe/Au devices treated with pH1 to pH4 are tabulated in Table 7.13. For these sets of experiments, the measurements were carried out at a frequency of 1.0 MHz, between the bias voltage range of -1.00 and 1.00 V at 300 K. The reported doping density N_D in this work was obtained using the Mott-Schottky plot as described in Sect. 3.5.2. The ε_r value was taken to be 11 [60], while the slope was obtained from the intercept of the Mott-Schottky plot as shown in Fig. 7.27b.

Using Equation 3.53 the effective density of states N_c was calculated to be 9.15×10^{17} cm^3, where h is Planck's constant, m_e^* is the effective electron mass, T is the temperature at 300 K, and k is Boltzmann's constant. The electron mobility $\mu_\perp$ was calculated using Equation 3.57.

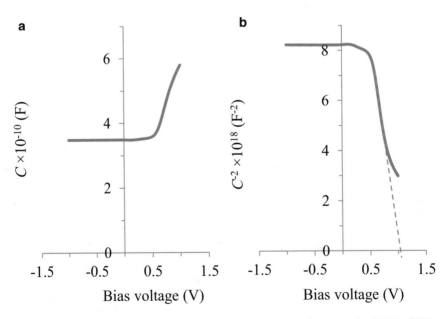

Fig. 7.27 A typical (**a**) capacitance-voltage and (**b**) Mott-Schottky plot of the glass/FTO/n-CdS/n-CdTe/p-CdTe/Au layer treated with pH2 CdCl$_2$

As observed in Fig. 7.27, the fabricated devices treated with $CdCl_2$ at pH2 were fully depleted at the reverse bias through the zero voltage and towards ~0.5 V in the forward bias with the depletion width W which exceeds the fabricated device thickness of ~1350 nm. Increasing the voltage in the forward bias to ~0.5 V and above, a gradual reduction in the depletion width was observed with a corresponding increase in capacitance. This observation was similar for glass/FTO/n-CdS/n-CdTe/p-CdTe/Au devices treated with $CdCl_2$ at pH3 and pH4. Furthermore, the calculated N_D (see Table 7.13) for the devices treated with $CdCl_2$ at pH2–pH4 is of the same order of magnitude (10^{14} cm^{-3}). High-efficiency solar cells have been reported to have N_D values within the (~1.0×10^{14}–5×10^{15} cm^{-3}) [61, 62]. The N_D value of the devices activated with $CdCl_2$ at pH1 signifies high doping which might result in a loss of J_{sc}, incorporation of defects within the crystal lattice, shrinkage of the depletion width and the consequent reduction in the photo-generated current collection efficiency [63]. These observations coupled with the high defect density (R&G) centres might be the cause of the reduction in the charge carrier mobility of the devices activated with pH1 $CdCl_2$ as compared to pH2–pH4 $CdCl_2$ treatment.

7.9.4 Summations

In conclusion, this experimental work has explored the effect of $CdCl_2$ post-growth treatment pH on both material and fabricated CdS/CdTe device properties. It was observed that better material and device properties could be achieved at pH2 $CdCl_2$ activation treatment. Although the device parameters such as the V_{oc} show no distinct difference after treatment with pH2 to pH4 $CdCl_2$, the $CdCl_2$ treatment at pH1 shows low material quality as observed in the structural, morphological and compositional properties, while the overall fabricated device efficiency was low.

7.10 Effect of the Inclusion of Gallium in the Normal CdCl₂ Treatment of CdS-/CdTe-Based Solar Cells

Post-growth treatment (PGT) has been considered as an integral part of achieving highly efficient solar cells [1]. This is justified by recrystallisation and grain growth, optimisation of electrical conductivity and doping concentration, passivation of grain boundaries, optimisation of the CdS/CdTe interface morphology, improvement in Cd/Te stoichiometric composition and reduction of Te precipitation, amongst other advantages observed after PGT in the presence of some halogen-based salts and gases [5, 30, 40, 64, 65]. As documented in the literature, it is challenging to avoid the presence of Te precipitates completely in CdTe by modifying the growth technique, growth process or post-growth treatment [66–68]. The inclusion of gallium in the normal $CdCl_2$ treatment of CdTe is due to its unique property of dissolving Te precipitates in CdTe material as suggested in the literature [6, 67].

7.10.1   Effect of the Inclusion of Gallium in the Normal CdCl₂ Treatment on the Material Properties of CdS-/CdTe-Based Solar Cell

7.10.1.1   Optical Absorption Analysis

Figure 7.28a shows the optical absorption curves, and Fig. 7.28b shows the plot of absorption edge slopes and bandgaps against the AD, CCT and GCT glass/FTO/n-CdS/n-CdTe/p-CdTe layers, while the numerical values are shown in Table 7.14. As observed in Fig. 7.28a, b, the bandgaps of all the AD, CCT and GCT layers fall within the CdTe bulk bandgap range of 1.45–1.50 eV [23], even for the as-deposited layers without any PGT to modify its optical parameters. Although there were no clear differences in the bandgap due to the material quality of the as-deposited CdTe layer, the difference in the absorption edge is clearly observed in Fig. 7.28b.

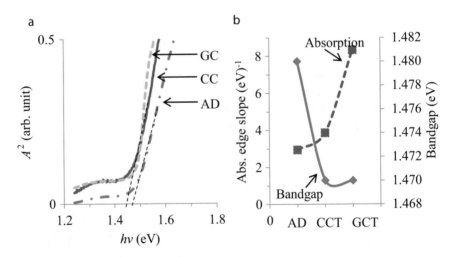

Fig. 7.28 Graphs of (**a**) square of absorption against photon energy and (**b**) optical bandgap and slope of absorption edge against treatment conditions for AD, CCT and GCT glass/FTO/n-CdS/n-CdTe/p-CdTe layers

Table 7.14 Summary of the effect of different treatments on bandgap and the slope of absorption edges

Sample	Bandgap (eV)	Absorption edge slope (eV^{-1})
AD	1.48	2.94
CCT	1.47	3.85
GCT	1.47	8.33

It is well documented in the literature that the sharpness of the absorption edge signifies superior semiconductor layer optical property based on lesser impurity energy levels and defects in the thin film [5, 23, 69]. Based on this submission, it could be interpreted that the layer with the superior optical quality is the GCT layer and the least is the as-deposited CdTe layer. It is interesting to observe an inverse relationship (see Fig. 7.28b) between the bandgap and the absorption edge slope which further buttresses the GCT glass/FTO/n-CdS/n-CdTe/p-CdTe material superiority.

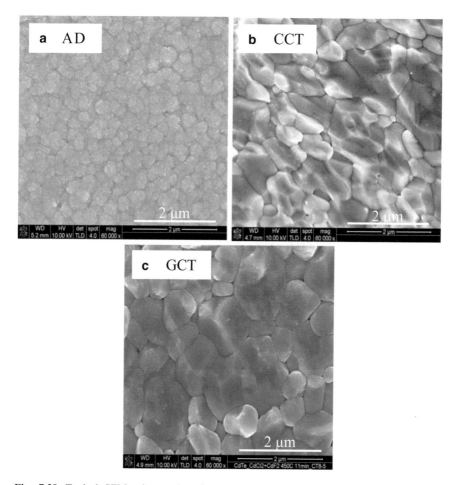

Fig. 7.29 Typical SEM micrographs of (**a**) AD, (**b**) CCT and (**c**) GCT-CdTe thin films electrodeposited on CdS layer at 1370 mV

7.10.1.2 Morphological Analysis

Figure 7.29a–c shows the morphology of AD, CCT and GCT glass/FTO/*n*-CdS/
n-CdTe/*p*-CdTe thin films. As observed in Fig. 7.29, the underlying substrates are
fully covered by the CdTe layer before and after different treatments.

The as-deposited glass/FTO/*n*-CdS/*n*-CdTe/*p*-CdTe layer shows agglomeration
of small crystallites to form cauliflower-like larger grains. After post-growth treatment such as CCT and GCT, grain growths within the range of 100–2000 nm and
200–2600 nm were observed, respectively. The influence of CdCl$_2$ in PGT of CdTe
has been explicitly explored in the literature as it affects the improvement of material
and device quality of CdTe-based solar cells [1, 5, 40]. Further improvements in
grain size can be observed in morphological property of the GCT-treated glass/FTO/
n-CdS/*n*-CdTe/*p*-CdTe layer as shown in Fig. 7.29c as compared with the CCT in
Fig. 7.29b in this work. This improvement signals the positive effect of the inclusion
of Ga in the usual CCT of CdTe.

7.10.1.3 Compositional Analysis

Figure 7.30 shows the atomic composition of a typical glass/FTO/*n*-CdS/*n*-CdTe/
p-CdTe multilayer configuration as detected by EDX technique. It should be noted
that in addition to Cd and Te, the presence of S, Si, Sn and F may also be observed
due to underlying glass/FTO/CdS substrate, Cl or Ga due to the PGT utilised and O
due to layer oxidation. As expected, a low Cd/Te composition ratio was observed for
the AD layer due to the Te richness during the growth of the top *p*-CdTe layer at
lower cathodic potential. This observation further bolsters the fact that provided
CdTe is not subjected to any extrinsic doping and the conductivity type of CdTe is
composition dependent for the as-deposited layer. A shift towards unity of the Cd/Te
ratio was observed after CCT and GCT treatments.

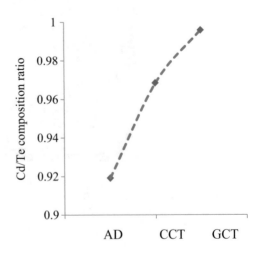

Fig. 7.30 Graph of
compositional ratio of
Cd/Te atoms in a glass/FTO/
n-CdS/*n*-CdTe/*p*-CdTe
configuration against
different post-growth
treatments

This observation has been reported in the literature [1, 5, 9, 40] as one of the advantages of PGT of CdTe. It should be noted that in addition to elemental composition, self-compensation and doping effect can take place during PGT. In the case of GCT, the presence of gallium can also remove Te precipitate and act as an *n*-type dopant during the heat treatment [6, 26]. Therefore, the combination of all these processes seems to produce beneficial properties for the CdTe layer and the glass/FTO/*n*-CdS/*n*-CdTe/*p*-CdTe structure.

7.10.1.4   Structural Analysis

Figure 7.31 shows the structural analysis of electrodeposited glass/FTO/*n*-CdS/ *n*-CdTe/*p*-CdTe layers under different conditions, and Table 7.15 summarises the results of X-ray diffraction (XRD) analysis on the effect of different post-growth treatments on glass/FTO/*n*-CdS/*n*-CdTe/*p*-CdTe layers. It was observed that no reflection could be attributed to the underlying CdS layer [70] except for the hexagonal CdS (002) which coincides with the FTO peak at angle $2\theta = 26.68°$ and cannot be ascertained. Other reflections attributed to FTO were observable at $2\theta = 32.9°$, $37.1°$ and $51.6°$. XRD reflections assigned to cubic (111), (220) and (311) CdTe phases at $2\theta = 23.88°$, $38.65°$ and $45.84°$ were also observed. It can be deduced from Fig. 7.31 that under all conditions explored in this work, the most

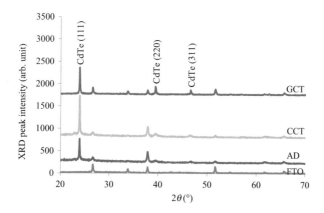

Fig. 7.31   XRD spectra of glass/FTO/*n*-CdS/*n*-CdTe/*p*-CdTe structures for different conditions (AD, CCT and GCT)

Table 7.15   The XRD analysis on the effect of PGT on CdTe layers

Treatment	2θ (°)	Lattice spacing (Å)	FWHM (°)	Crys. size D (nm)	Plane (*hkl*)	Assignments
AD	23.94	3.72	0.162	52.3	(111)	Cubic
CCT	23.92	3.72	0.162	52.3	(111)	Cubic
GCT	23.92	3.72	0.162	52.3	(111)	Cubic

intense XRD reflection is observed at $2\theta = 23.85°$. For the CCT glass/FTO/n-CdS/n-CdTe/p-CdTe layer, an increase in the cubic (111)C orientation was observed as compared to the AD without any observable change in the other CdTe reflections. However, for GCT glass/FTO/n-CdS/n-CdTe/p-CdTe layer, randomisation of crystallite orientation was observed. This is usually seen with the collapse of (111) peak and increase in (220) and (311) peaks. As reported recently [71], these changes suddenly occur when the grain boundaries are melted due to the presence of impurities such as excess Cd, Cl and O. The presence of Ga seems to enhance the decrease in the (111) peak and increase the (220) and (311) peak intensities due to increased randomisation of crystal orientations.

From the comparison between the AD, CCT and GCT glass/FTO/n-CdS/n-CdTe/p-CdTe layers, it could be inferred that the inclusion of gallium in the normal cadmium chloride treatment PGT may have triggered the recrystallisation and reorientation of the crystalline planes in this work. It should be noted that alterations in XRD patterns also depend on the underlying substrates and heat treatment conditions used [71].

As shown in Table 7.15, there was no clear distinction between the glass/FTO/n-CdS/n-CdTe/p-CdTe layers explored in this work as concerning the full width at half maximum (FWHM), lattice spacing and crystallite size calculated using Scherrer's equation. This observation might be due to the limitations of the use of Scherrer's equation [72] or the XRD analysis software on polycrystalline material layers with large crystals.

7.10.1.5 Photoelectrochemical (PEC) Cell Study

Table 7.16 shows the PEC signal of glass/FTO/p-CdTe layer after different PGTs. It was observed that the p-conduction type of the as-deposited CdTe layer was retained after different treatments. Although, a shift in the PEC signal towards the n-type conduction region was also observable in both the CCT and GCT p-CdTe layers. Both Cl and Ga in smaller concentrations act as n-type dopants in CdTe. The presence of both Cl and Ga in PGT shows the reduction of the p-type nature of CdTe due to the introduction of n-dopants.

This observation depicts the movement of the FL which was close to the valence band towards the middle of the bandgap due to the alteration in doping as a result of the heat treatment condition [40] and Cd/Te compositional changes [73], amongst other factors. Other incorporated layers such as n-CdS and n-CdTe have been known to retain their conductivity type with a slight shift towards the opposite conductivity type [9, 33].

Table 7.16 PEC cell measurements on p-CdTe layers after different treatments

Treatment	V_L (mV)	V_D (mV)	$(V_L - V_D)$ (mV)	Conduction type
AD	−61	−84	23	P
CCT	−78	−96	18	p
GCT	−42	−47	5	p

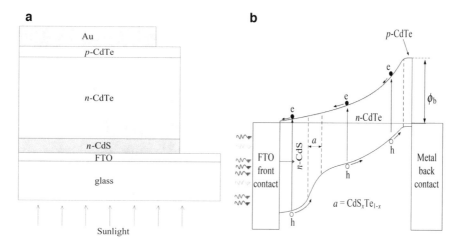

Fig. 7.32 (**a**) Schematic diagram and (**b**) the band diagram of the glass/FTO/*n*-CdS/*n*-CdTe/*p*-CdTe/Au thin-film solar cell

7.10.2   The Effect of the Inclusion of Gallium in the Normal CdCl₂ Treatment on Solar Cell Device Parameters

Based on the analysis as discussed in Sect. 7.10.1, the glass/FTO/*n*-CdS/*n*-CdTe/*p*-CdTe layer schematics and the band diagram can be represented by Fig. 7.32a, b, respectively.

Figure 7.33a–c show the current-voltage (*I-V*) curves of glass/FTO/*n*-CdS/*n*-CdTe/*p*-CdTe/Au layers with different PGT conditions, measured under AM1.5 as discussed in Sect. 3.5.1.2. From these *I-V* curves measured under the illuminated AM1.5 condition, solar cell parameters can be determined. Experimentally observed solar cell parameters for the different conditions are summarised in Table 7.17 for three champion cells. Both the series resistance R_s and shunt resistance R_{sh} were calculated from the inverse slopes of the linear-linear *I-V* curve in the forward and reverse bias, respectively, under AM1.5 illuminated conditions.

The high R_s and low R_{sh} values as observed in the fabricated solar cell incorporating AD-CdTe can be directly attributed to low semiconductor material quality as described by Soga (2004) [19]. It is clearly observed in Fig. 7.33 and Table 7.17 that the improvement in the electrical properties of the glass/FTO/*n*-CdS/*n*-CdTe/*p*-CdTe/Au devices can be achieved after some chlorine-based treatment [74].

The addition of gallium into the regular CdCl₂ treatment further enhances the device performance. This enhancement can be attributed to the material properties' improvement after treatment as discussed in Sect. 7.10.1. It should be noted that the short-circuit current density as observed in this work is higher than the Shockley-Queisser limit for a single *p-n* junction [11] as a result of the incorporation of the multilayer *n-n-p* configuration [12]. The structure represents an early stage of the graded bandgap, multilayer device configuration.

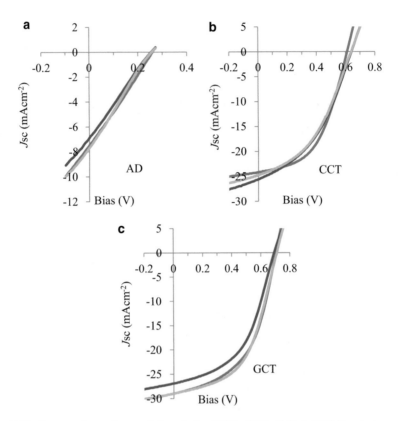

Fig. 7.33 Current-voltage characteristics of glass/FTO/*n*-CdS/*n*-CdTe/p-CdTe/Au devices with AD, CCT and GCT conditions

Table 7.17 Tabulated device parameters obtained from *I-V* measurements under AM1.5 illuminated condition

Treatment	R_s (Ω)	R_{sh} (kΩ)	J_{sc} (mAcm^{-2})	V_{oc} (V)	FF	Efficiency (%)
AD	1027	1.4	7.1	0.25	0.24	0.43
	1062	1.5	7.8	0.26	0.24	0.49
	1180	1.3	7.6	0.25	0.24	0.46
CCT	302	6.4	24.2	0.61	0.50	7.50
	353	3.8	25.5	0.64	0.43	7.01
	354	3.2	24.8	0.64	0.43	6.84
GCT	273	8.0	29.9	0.72	0.52	11.21
	319	7.1	27.4	0.70	0.52	9.97
	276	8.0	29.3	0.73	0.52	11.12

Furthermore, the comparatively higher FF, J_{sc} and V_{oc} observed with cells fabricated using GCT-CdTe can be attributed to the incorporation of *n*-dopant treatment such as gallium in CdTe by the introduction of excess electrons into the crystal lattice to boost conductivity and also further improvement in the material quality as observed in Sect. 7.10.1. Similar observations have been recorded in the literature with the incorporation of *n*-dopant to CdTe [41, 75, 76]. Although the effect of the incorporation of gallium on bandgap defects cannot be depicted from the *I-V* results, the improvement in the overall electronic properties of the fabricated devices is observable.

7.10.3 Summations

The effect of the inclusion of Ga to the regular $CdCl_2$ post-growth treatment on the material and electronic properties of CdS/CdTe-based layers has been explored in this work. The optical analysis shows that the grown CdTe layer is within the standard bulk CdTe bandgap range with material superiority after $CdCl_2$/Ga treatment due to the steeper absorption edge slope—lesser impurity energy levels and defects in the thin film. The morphological studies show full material coverage and grain growth after both CCT and GCT. The compositional analysis shows the improvement of stoichiometry when treated with GCT. The structural analysis shows improvement in the XRD peak intensity reflection of the as-deposited glass/FTO/*n*-CdS/*n*-CdTe/*p*-CdTe after both CCT and GCT with the preferred orientation along the cubic (111) plane. A more pronounced recrystallisation was observed after GCT with a comparative reduction in the (111)C peak and an increase in the (220)C reflection showing enhanced recrystallisation. Improvement in the electrical properties of the fabricated glass/FTO/*n*-CdS/*n*-CdTe/*p*-CdTe/Au was observed after PGT with GCT showing better results than CCT owing to the gallium inclusion in the treatment. With further material optimisation of gallium doping in the CdTe treatment, improved electronic properties can be achieved. Work is ongoing on the optimisation of Ga concentration and treatment parameters required for this inclusion.

7.11 Summary of the Effect of Gallium Chloride Treatment pH

Similar to the experimental processes as discussed in Sect. 7.9, the effects of the pH of $CdCl_2 + Ga_2(SO_4)_3$ (or GCT) on glass/FTO/*n*-CdS/*n*-CdTe/*p*-CdTe/Au devices were explored and are summarised in this section. Due to brevity and the significance of the importance of the study, only pH1 and pH2 data are presented.

Figure 7.34 shows the typical SEM, EDX, optical absorption and XRD peak patterns for CdTe post-growth treated with pH1 and pH2 GCT. The *I-V* and *C-V* characteristic properties of typical devices which have undergone GCT of pH1 and pH2 are presented in Table 7.18.

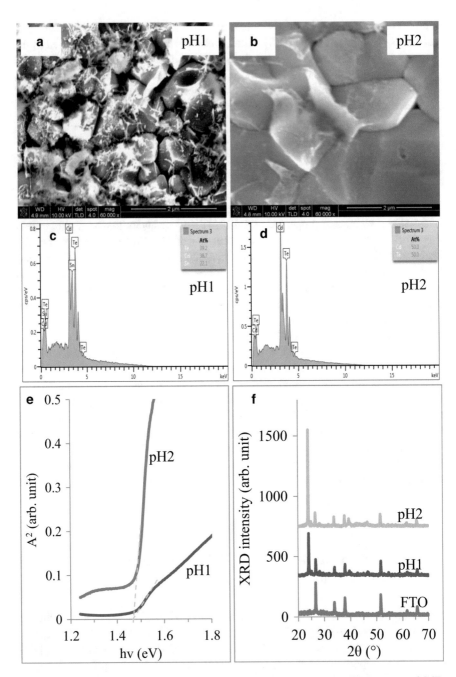

Fig. 7.34 Typical SEM of CdTe after GCT at (**a**) pH1 and (**b**) pH2. Typical EDX spectra of CdTe after GCT at (**c**) pH1 and (**d**) pH2. (**e**) Typical absorption curve of CdTe treated with GCT at pH1 and pH2 and (**f**) typical XRD peaks of CdTe treated with GCT at pH1 and pH2

Table 7.18 Summary of device parameters obtained from *I-V* (both under illuminated and dark conditions) and *C-V* (dark condition) for glass/FTO/*n*-CdS/*n*-CdTe/*p*-CdTe/Au solar cells treated with pH1 and pH2 CdCl$_2$ + Ga$_2$(SO$_4$)$_3$

GCT pH	pH1	pH2
I-V under dark condition		
R_{sh} (Ω)	9231	$>10^6$
R_s (kΩ)	1.25	0.66
log (RF)	0.60	4.00
I_o (A)	3.16×10^{-6}	1.25×10^{-9}
n	>2.00	1.67
ϕ_b (eV)	0.56	>0.81
I-V under AM1.5 illumination condition		
I_{sc} (mA)	0.63	0.96
J_{sc} (mAcm^{-2})	20.06	30.57
V_{oc} (V)	0.34	0.71
Fill factor	0.31	0.56
Efficiency (%)	2.11	12.16
C-V under dark condition		
$\sigma \times 10^{-4}$ (Ω.cm)$^{-1}$	11.39	0.36
N_D (cm^{-3})	6.22×10^{16}	1.12×10^{15}
$\mu_{\perp}$ (cm^{2}V^{-1} s^{-1})	0.11	20.06
C_o (pF)	150	350
W (nm)	203.9	873.8

As shown in the SEM micrographs (see Fig. 7.34a, b), an erosion of the CdTe layer morphology was observable at pH1, while the morphology of the CdTe treated at pH2 appears intact with low pinhole density and good coverage of the underlying layers. It should be noted that the eroded CdTe surface at pH1 is due to high acidity (low pH) and the dissolution of Cd whereby making the CdTe surface Te-rich [7] as observed in Fig. 7.34c, d. Furthermore, due to the thinness of the films (within nanoscale), the nucleation mechanism of electrodeposited materials and the columnar growth nature of the electroplated materials, the erosion of the CdTe surface will result in opening pores along the grain boundaries which may result into shunt paths for charge carriers. The optical absorbance measurements (Fig. 7.34e) show that both the CdTe layers treated with pH1 and pH2 GCT have bandgaps close to the 1.45 eV, standard bandgap for bulk CdTe. The deviation of the bandgap away from the standard and the reduction in the absorption edge is due to compositional alteration away from stoichiometry [5, 77]. This observation is further corroborated with higher XRD peak intensity as shown in Fig. 7.34f which signifies higher crystallinity for the pH2-treated sample.

With reference to Table 7.18, the effect of material deterioration due to the harsh erosion at GCT pH1 is summed up by the low R_{sh} [19]. Furthermore, lower conversion efficiency was observed for the pH1 GCT devices which might be due to the low R_{sh} which creates alternative paths for charge carriers and reduction in both the current density (J_{sc}) and charge carrier mobility ($\mu_{\perp}$) as it is observed in the pH1 GCT devices.

7.12 Conclusions

Solar cells of different configurations, material layers, thicknesses and post-growth treatment conditions were successfully fabricated and explored using both I-V and C-V techniques. The results were systematically presented starting with the optimisation of the n-CdS window layer thickness. Based on observed electronic parameters, the optimised thickness of n-CdS in all-electrodeposited glass/FTO/n-CdS/n-CdTe/p-CdTe/Au configuration was determined to be between 100 and 150 nm.

The optimised n-CdS thickness was utilised as a substrate for CdTe grown at different cathodic voltages around the cathodic voltage in which stoichiometric CdTe layers were grown. This resulted into either glass/FTO/n-CdS/p-CdTe/Au, n-p junctions or glass/FTO/n-CdS/n-CdTe/Au, n-n + SB configurations in which better electronic parameters were observed in the n-n + SB architecture. In a view to further reduce CdS thickness, n-ZnS of 50 nm and CdS of 65 nm were incorporated in glass/FTO/n-ZnS/n-CdS/n-CdTe/Au structures and compared with the base glass/FTO/n-CdS/n-CdTe/Au. Higher efficiency and improvement in other parameters were obtained for the glass/FTO/n-ZnS/n-CdS/n-CdTe/Au owing to the multilayer configuration based on the advantages of graded bandgap configuration. Further to this, the effects of F-, Cl-, I- and Ga-doped CdTe to give n-CdTe incorporated into the glass/FTO/n-CdS/n-CdTe/p-CdTe/Au configuration were also explored. Better prospects for achieving high efficiency were recorded for the devices incorporating F-, Cl- and most especially the Ga-doped n-CdTe layers.

Also explored in this chapter is the effect of the incorporation of Ga in the normal $CdCl_2$ post-growth treatment and the effect of pH in both the CCT and GCT treatments of the glass/FTO/n-CdS/n-CdTe/p-CdTe/Au configuration. Further to this, the effect of p-CdTe in the glass/FTO/n-CdS/n-CdTe/p-CdTe/Au configuration was also explored with the optimised thickness of ~35 nm. In addition, the variation as a function of batch and stability as a function of time were also explored. Conclusively, both the n-n-p and the n-n-n + SB multilayer architectures show great prospects for further exploration and development. In this research work, the highest efficiencies obtained in the glass/FTO/n-CdS/p-CdTe/Au (an n-p junction) and glass/FTO/n-CdS/n-CdTe/Au (an n-n + SB) are 4.5% and 7.5%, respectively. The highest efficiencies for the glass/FTO/n-CdS/n-CdTe/p-CdTe/Au and glass/FTO/n-CdS/n-CdTe/p-CdTe/Cu-Au are 15.3% and 18.5%, respectively, while the highest conversion efficiency obtained for the glass/FTO/n-ZnS/n-CdS/n-CdTe/Au configuration was 14.1%.

References

1. I.M. Dharmadasa, Review of the CdCl2 treatment used in CdS/CdTe thin-film solar cell development and new evidence towards improved understanding. Coatings **4**, 282–307 (2014). https://doi.org/10.3390/coatings4020282

2. O.I. Olusola, *Optoelectronic Devices Based on Graded Bandgap Structures Utilising Electroplated Semiconductors* (Sheffield Hallam University, Sheffield, 2016)
3. H.I. Salim, *Multilayer Solar Cells Based on CdTe Grown from Nitrate Precursor* (Sheffield Hallam University, Sheffield, 2016)
4. O.K. Echendu, *Thin-Film Solar Cells Using All-Electrodeposited ZnS, CdS and CdTe Materials* (Sheffield Hallam University, Sheffield, 2014)
5. A. Bosio, N. Romeo, S. Mazzamuto, V. Canevari, Polycrystalline CdTe thin-films for photovoltaic applications. Prog. Cryst. Growth Charact. Mater. **52**, 247–279 (2006). https://doi.org/10.1016/j.pcrysgrow.2006.09.001
6. P. Fernández, Defect structure and luminescence properties of CdTe based compounds. J. Optoelectron. Adv. Mater. **5**, 369–388 (2003)
7. I.M. Dharmadasa, Recent developments and progress on electrical contacts to CdTe, CdS and ZnSe with special reference to barrier contacts to CdTe. Prog. Cryst. Growth Charact. Mater. **36**, 249–290 (1998). https://doi.org/10.1016/S0960-8974(98)00010-2
8. I.M. Dharmadasa, C.J. Blomfield, C.G. Scott, R. Coratger, F. Ajustron, J. Beauvillain, Metal/n-CdTe interfaces: a study of electrical contacts by deep level transient spectroscopy and ballistic electron emission microscopy. Solid State Electron. **42**, 595–604 (1998). https://doi.org/10.1016/S0038-1101(97)00296-7
9. H.I. Salim, V. Patel, A. Abbas, J.M. Walls, I.M. Dharmadasa, Electrodeposition of CdTe thin-films using nitrate precursor for applications in solar cells, J. Mater. Sci. Mater. Electron. 26 (2015) 3119–3128. doi: https://doi.org/10.1007/s10854-015-2805-x.
10. J.E. Granata, J.R. Sites, Effect of CdS thickness on CdS/CdTe quantum efficiency, in *Conf. Rec. Twenty Fifth IEEE Photovolt. Spec. Conf. 1996* (2000), pp. 853–856. https://doi.org/10.1109/PVSC.1996.564262
11. W. Shockley, H.J. Queisser, Detailed balance limit of efficiency of p-n junction solar cells. J. Appl. Phys. **32**, 510 (1961). https://doi.org/10.1063/1.1736034
12. A. De Vos, Detailed balance limit of the efficiency of tandem solar cells. J. Phys. D. Appl. Phys. **13**, 839–846 (2000). https://doi.org/10.1088/0022-3727/13/5/018
13. J.S. Lee, Y.K. Jun, H.B. Im, Effects of CdS film thickness on the photovoltaic properties of sintered CdS / CdTe solar cells. J. Electrochem. Soc. **134**, 248–251 (1987). https://doi.org/10.1149/1.2100417
14. S.G. Kumar, K.S.R.K. Rao, Physics and chemistry of CdTe/CdS thin-film heterojunction photovoltaic devices: fundamental and critical aspects. Energy Environ. Sci. **7**, 45–102 (2014). https://doi.org/10.1039/C3EE41981A
15. I.M. Dharmadasa, *Advances in Thin-Film Solar Cells* (Pan Stanford, Singapore, 2013)
16. I.M. Dharmadasa, J.D. Bunning, A.P. Samantilleke, T. Shen, Effects of multi-defects at metal/semiconductor interfaces on electrical properties and their influence on stability and lifetime of thin-film solar cells. Sol. Energy Mater. Sol. Cells **86**, 373–384 (2005). https://doi.org/10.1016/j.solmat.2004.08.009
17. I.M. Dharmadasa, Third generation multi-layer tandem solar cells for achieving high conversion efficiencies. Sol. Energy Mater. Sol. Cells **85**, 293–300 (2005). https://doi.org/10.1016/j.solmat.2004.08.008
18. I.M. Dharmadasa, A.P. Samantilleke, N.B. Chaure, J. Young, New ways of developing glass/conducting glass/CdS/CdTe/metal thin-film solar cells based on a new model. Semicond. Sci. Technol. **17**, 1238–1248 (2002). https://doi.org/10.1088/0268-1242/17/12/306
19. T. Soga, *Nanostructured Materials for Solar Energy Conversion* (Elsevier Science, 2006), p. 614. https://www.elsevier.com/books/nanostructured-materials-for-solar-energy-conversion/soga/978-0-444-52844-5
20. C. Ni, P. Shah, A.M. Sarangan, Effects of different wetting layers on the growth of smooth ultra-thin silver thin-films, in ed. by E.M. Campo, E.A. Dobisz, L.A. Eldada (2014), p. 91700L. https://doi.org/10.1117/12.2061256
21. T. Yasuda, K. Hara, H. Kukimoto, Low resistivity Al-doped ZnS grown by MOVPE. J. Cryst. Growth **77**, 485–489 (1986). https://doi.org/10.1016/0022-0248(86)90341-6

22. M.L. Madugu, O.I.-O. Olusola, O.K. Echendu, B. Kadem, I.M. Dharmadasa, Intrinsic doping in electrodeposited ZnS thin-films for application in large-area optoelectronic devices. J. Electron. Mater. **45**, 2710–2717 (2016). https://doi.org/10.1007/s11664-015-4310-7

23. T.L. Chu, S.S. Chu, Thin-film II–VI photovoltaics. Solid State Electron. **38**, 533–549 (1995). https://doi.org/10.1016/0038-1101(94)00203-R

24. O. Echendu, I. Dharmadasa, Graded-bandgap solar cells using all-electrodeposited ZnS, CdS and CdTe thin-films. Energies **8**, 4416–4435 (2015). https://doi.org/10.3390/en8054416

25. X. Liu, Y. Jiang, F. Fu, W. Guo, W. Huang, L. Li, Facile synthesis of high-quality ZnS, CdS, CdZnS, and CdZnS/ZnS core/shell quantum dots: characterization and diffusion mechanism. Mater. Sci. Semicond. Process. **16**, 1723–1729 (2013). https://doi.org/10.1016/j.mssp.2013.06. 007

26. J.M. Woodcock, A.K. Turner, M.E. Ozsan, J.G. Summers, Thin-film solar cells based on electrodeposited CdTe, in *Conf. Rec. Twenty-Second IEEE Photovolt. Spec. Conf.—1991, IEEE* (1991), pp. 842–847. https://doi.org/10.1109/PVSC.1991.169328

27. K. Zanio, *Semiconductors and Semimetals* (Academic, New York, 1978). http://shu.summon. serialssolutions.com/2.0.0/link/0/eLvHCXMwdV3JCsIwEB1cEAQPrrgV-gNKmyZNPYvFu 94l6bQ3K1j_HydDXXA5Zg7DJJB5me0FIBLrYPXhE8LEUZ8lRggjsgBlgBuptSqw0Chzrs y80Rg848ZXCuObQZ_iCKmCyN3HJjQJON2LqOaiYzdM7pnQiml0hCaUCiNVM-481sn7l wYMKGkfWm7IYACNvBxCh9sws2o

28. I.M. Dharmadasa, A.B. McLean, M.H. Patterson, R.H. Williams, Schottky barriers and interface reactions on chemically etched n-CdTe single crystals. Semicond. Sci. Technol. **2**, 404–412 (1987). https://doi.org/10.1088/0268-1242/2/7/003

29. S. Tanaka, J.A. Bruce, M.S. Wrighton, Deliberate modification of the behavior of n-type cadmium telluride/electrolyte interfaces by surface etching. Removal of Fermi level pinning. J. Phys. Chem. **85**, 3778–3787 (1981). https://doi.org/10.1021/j150625a015

30. I.M. Dharmadasa, O.K. Echendu, F. Fauzi, N.A. Abdul-Manaf, O.I. Olusola, H.I. Salim, M.L. Madugu, A.A. Ojo, Improvement of composition of CdTe thin-films during heat treatment in the presence of CdCl2. J. Mater. Sci. Mater. Electron. **28**, 2343–2352 (2017). https://doi.org/ 10.1007/s10854-016-5802-9

31. J. Britt, C. Ferekides, Thin-film CdS/CdTe solar cell with 15.8% efficiency. Appl. Phys. Lett. **62**, 2851–2852 (1993). https://doi.org/10.1063/1.109629

32. T. Potlog, L. Ghimpu, P. Gashin, A. Pudov, T. Nagle, J. Sites, Influence of annealing in different chlorides on the photovoltaic parameters of CdS/CdTe solar cells. Sol. Energy Mater. Sol. Cells **80**, 327–334 (2003). https://doi.org/10.1016/j.solmat.2003.08.007

33. N.A. Abdul-Manaf, A.R. Weerasinghe, O.K. Echendu, I.M. Dharmadasa, Electro-plating and characterisation of cadmium sulphide thin-films using ammonium thiosulphate as the sulphur source. J. Mater. Sci. Mater. Electron. **26**, 2418–2429 (2015). https://doi.org/10.1007/s10854-015-2700-5

34. B.E. McCandless, K.D. Dobson, Processing options for CdTe thin-film solar cells. Sol. Energy **77**, 839–856 (2004). https://doi.org/10.1016/j.solener.2004.04.012

35. D.W. Lane, A review of the optical band gap of thin-film CdSxTe 1-x. Sol. Energy Mater. Sol. Cells **90**, 1169–1175 (2006). https://doi.org/10.1016/j.solmat.2005.07.003

36. D.A. Wood, K.D. Rogers, D.W. Lane, D.A. Wood, K.D. Rogers, J.A. Coath, Optical and structural characterization of CdS x Te 1- x thin-films for solar cell applications. J. Phys. Condens. Matter **12**, 4433–4450 (2000). https://doi.org/10.1088/0953-8984/12/19/312 http:// stacks.iop.org/0953-8984/12/i=19/a=312.

37. E.Q.B. Macabebe, E.E. van Dyk, Parameter extraction from dark current–voltage characteristics of solar cells. S. Afr. J. Sci. **104**, 401–404 (2008). http://www.scielo.org.za/scielo.php? script=sci_arttext&pid=S0038-23532008000500017

38. S.M. Sze, K.K. Ng, *Physics of Semiconductor Devices* (Wiley, Hoboken, 2006). https://doi.org/ 10.1002/0470068329

39. K. Masuko, M. Shigematsu, T. Hashiguchi, D. Fujishima, M. Kai, N. Yoshimura, T. Yamaguchi, Y. Ichihashi, T. Mishima, N. Matsubara, T. Yamanishi, T. Takahama,

M. Taguchi, E. Maruyama, S. Okamoto, Achievement of more than 25% conversion efficiency with crystalline silicon heterojunction solar cell. IEEE J. Photovoltaics **4**, 1433–1435 (2014). https://doi.org/10.1109/JPHOTOV.2014.2352151

40. B.M. Basol, Processing high efficiency CdTe solar cells. Int. J. Sol. Energy **12**, 25–35 (1992). https://doi.org/10.1080/01425919208909748

41. S. Mazzamuto, L. Vaillant, A. Bosio, N. Romeo, N. Armani, G. Salviati, A study of the CdTe treatment with a Freon gas such as CHF2Cl. Thin Solid Films **516**, 7079–7083 (2008). https://doi.org/10.1016/j.tsf.2007.12.124

42. J.D. Major, L. Bowen, R.E. Treharne, L.J. Phillips, K. Durose, NH 4 Cl alternative to the CdCl 2 treatment step for CdTe thin-film solar cells. IEEE J. **5**, 386–389 (2015). https://doi.org/10.1109/JPHOTOV.2014.2362296

43. B. Maniscalco, A. Abbas, J.W. Bowers, P.M. Kaminski, K. Bass, G. West, J.M. Walls, The activation of thin-film CdTe solar cells using alternative chlorine containing compounds. Thin Solid Films **582**, 115–119 (2015). https://doi.org/10.1016/j.tsf.2014.10.059

44. B.E. McCandless, I. Youm, R.W. Birkmire, Optimization of vapor post-deposition processing for evaporated CdS/CdTe solar cells. Prog. Photovolt. Res. Appl. **7**, 21–30 (1999). https://doi.org/10.1002/(SICI)1099-159X(199901/02)7:1<21::AID-PIP244>3.0.CO;2-D

45. H. Bayhan, C. Ercelebi, Effects of post deposition treatments on vacuum evaporated CdTe thin-films and CdS/CdTe heterojunction devices. Turk. J. Phys. **22**, 441–451 (1998). http://www.scopus.com/scopus/inward/record.url?eid=2-s2.0-0347537252&partnerID=40&rel=R6.5.0

46. I.M. Dharmadasa, J.M. Thornton, R.H. Williams, Effects of surface treatments on Schottky barrier formation at metal/n-type CdTe contacts. Appl. Phys. Lett. **54**, 137 (1989). https://doi.org/10.1063/1.101208

47. V. Krishnakumar, J. Han, A. Klein, W. Jaegermann, CdTe thin-film solar cells with reduced CdS film thickness. Thin Solid Films **519**, 7138–7141 (2011). https://doi.org/10.1016/j.tsf.2010.12.118

48. G. Carotenuto, M. Palomba, S. De Nicola, G. Ambrosone, U. Coscia, Structural and photoconductivity properties of tellurium/PMMA films. Nanoscale Res. Lett. **10**, 1007 (2015). https://doi.org/10.1186/s11671-015-1007-z

49. B. Abad, M. Rull-Bravo, S.L. Hodson, X. Xu, M. Martin-Gonzalez, Thermoelectric properties of electrodeposited tellurium films and the sodium lignosulfonate effect. Electrochim. Acta **169**, 37–45 (2015). https://doi.org/10.1016/j.electacta.2015.04.063

50. S. Chun, S. Lee, Y. Jung, J.S. Bae, J. Kim, D. Kim, Wet chemical etched CdTe thin-film solar cells. Curr. Appl. Phys. **13**, 211–216 (2013). https://doi.org/10.1016/j.cap.2012.07.015

51. Z.H. Chen, C.P. Liu, H.E. Wang, Y.B. Tang, Z.T. Liu, W.J. Zhang, S.T. Lee, J.A. Zapien, I. Bello, Electronic structure at the interfaces of vertically aligned zinc oxide nanowires and sensitizing layers in photochemical solar cells. J. Phys. D. Appl. Phys. **44**, 325108 (2011). https://doi.org/10.1088/0022-3727/44/32/325108

52. Y. Shan, J.-J. Xu, H.-Y. Chen, Enhanced electrochemiluminescence quenching of CdS:Mn nanocrystals by CdTe QDs-doped silica nanoparticles for ultrasensitive detection of thrombin. Nanoscale **3**, 2916 (2011). https://doi.org/10.1039/c1nr10175g

53. N.A. Abdul-Manaf, H.I. Salim, M.L. Madugu, O.I. Olusola, I.M. Dharmadasa, Electro-plating and characterisation of CdTe thin-films using CdCl2 as the cadmium source. Energies **8**, 10883–10903 (2015). https://doi.org/10.3390/en81010883

54. P.J. Sellin, A.W. Dazvies, A. Lohstroh, M.E. Özsan, J. Parkin, Drift mobility and mobility-lifetime products in CdTe:Cl grown by the travelling heater method. IEEE Trans. Nucl. Sci. **52**, 3074–3078 (2005). https://doi.org/10.1109/TNS.2005.855641

55. J. Verschraegen, M. Burgelman, J. Penndorf, Temperature dependence of the diode ideality factor in CuInS2-on-Cu-tape solar cells. Thin Solid Films **480–481**, 307–311 (2005). https://doi.org/10.1016/j.tsf.2004.11.006

56. R.B. Godfrey, M.A. Green, Enhancement of MIS solar-cell "efficiency" by peripheral collection. Appl. Phys. Lett. **31**, 705–707 (1977). https://doi.org/10.1063/1.89487

57. I.M. Dharmadasa, A.A. Ojo, H.I. Salim, R. Dharmadasa, Next generation solar cells based on graded bandgap device structures utilising rod-type nano-materials. Energies **8**, 5440–5458 (2015). https://doi.org/10.3390/en8065440

58. D. Congreve, J. Lee, N. Thompson, E. Hontz, External quantum efficiency above 100% in a singlet-exciton-fission–based organic photovoltaic cell. Science **340**, 334–337 (2013). https://doi.org/10.1126/science.1232994

59. N.J.L.K. Davis, M.L. Bohm, M. Tabachnyk, F. Wisnivesky-Rocca-Rivarola, T.C. Jellicoe, C. Ducati, B. Ehrler, N.C. Greenham, Multiple-exciton generation in lead selenide nanorod solar cells with external quantum efficiencies exceeding 120%. Nat. Commun. **6**, 81–87 (2015). https://doi.org/10.1007/s13398-014-0173-7.2

60. I. Strzalkowski, S. Joshi, C.R. Crowell, Dielectric constant and its temperature dependence for GaAs, CdTe, and ZnSe. Appl. Phys. Lett. **28**, 350–352 (1976). https://doi.org/10.1063/1.88755

61. B.M. Basol, B. McCandless, Brief review of cadmium telluride-based photovoltaic technologies. J. Photonics Energy. **4**, 40996 (2014). https://doi.org/10.1117/1.JPE.4.040996

62. M. Gloeckler, I. Sankin, Z. Zhao, CdTe solar cells at the threshold to 20% efficiency. IEEE J. Photovoltaics **3**, 1389–1393 (2013). https://doi.org/10.1109/JPHOTOV.2013.2278661

63. T.J. Coutts, S. Naseem, High efficiency indium tin oxide/indium phosphide solar cells. Appl. Phys. Lett. **46**, 164–166 (1985). https://doi.org/10.1063/1.95723

64. H. Liu, Y. Tian, Y. Zhang, K. Gao, K. Lu, R. Wu, D. Qin, H. Wu, Z. Peng, L. Hou, W. Huang, Solution processed CdTe/CdSe nanocrystal solar cells with more than 5.5% efficiency by using an inverted device structure. J. Mater. Chem. C **3**, 4227–4234 (2015). https://doi.org/10.1039/C4TC02816C

65. H. Xue, R. Wu, Y. Xie, Q. Tan, D. Qin, H. Wu, W. Huang, Recent progress on solution-processed CdTe nanocrystals solar cells. Appl. Sci. **6**, 197 (2016). https://doi.org/10.3390/app6070197

66. I.M. Dharmadasa, O.K. Echendu, F. Fauzi, N.A. Abdul-Manaf, H.I. Salim, T. Druffel, R. Dharmadasa, B. Lavery, Effects of CdCl2 treatment on deep levels in CdTe and their implications on thin-film solar cells: a comprehensive photoluminescence study. J. Mater. Sci. Mater. Electron. **26**, 4571–4583 (2015). https://doi.org/10.1007/s10854-015-3090-4

67. N.V.V. Sochinskii, V.N.N. Babentsov, N.I.I. Tarbaev, M.D. Serrano, E. Dieguez, The low temperature annealing of p-cadmium telluride in gallium-bath. Mater. Res. Bull. **28**, 1061–1066 (1993). https://doi.org/10.1016/0025-5408(93)90144-3

68. J.C. Tranchart, P. Bach, A gas bearing system for the growth of CdTe. J. Cryst. Growth **32**, 8–12 (1976). https://doi.org/10.1016/0022-0248(76)90003-8

69. J. Han, C. Spanheimer, G. Haindl, G. Fu, V. Krishnakumar, J. Schaffner, C. Fan, K. Zhao, A. Klein, W. Jaegermann, Optimized chemical bath deposited CdS layers for the improvement of CdTe solar cells. Sol. Energy Mater. Sol. Cells **95**, 816–820 (2011). https://doi.org/10.1016/j.solmat.2010.10.027

70. T. Toyama, K. Matsune, H. Oda, M. Ohta, H. Okamoto, X-ray diffraction study of CdS/CdTe heterostructure for thin-film solar cell: influence of CdS grain size on subsequent growth of (111)-oriented CdTe film. J. Phys. D. Appl. Phys. **39**, 1537–1542 (2006). https://doi.org/10.1088/0022-3727/39/8/013

71. I.M. Dharmadasa, P. Bingham, O.K. Echendu, H.I. Salim, T. Druffel, R. Dharmadasa, G. Sumanasekera, R. Dharmasena, M.B. Dergacheva, K. Mit, K. Urazov, L. Bowen, M. Walls, A. Abbas, Fabrication of CdS/CdTe-based thin-film solar cells using an electrochemical technique. Coatings **4**, 380–415 (2014). https://doi.org/10.3390/coatings4030380

72. A. Monshi, Modified Scherrer equation to estimate more accurately nano-crystallite size using XRD. World J. Nano Sci. Eng. **2**, 154–160 (2012). https://doi.org/10.4236/wjnse.2012.23020

73. T.M. Razykov, N. Amin, B. Ergashev, C.S. Ferekides, D.Y. Goswami, M.K. Hakkulov, K.M. Kouchkarov, K. Sopian, M.Y. Sulaiman, M. Alghoul, H.S. Ullal, Effect of CdCl2 treatment on physical properties of CdTe films with different compositions fabricated by chemical molecular beam deposition. Appl. Sol. Energy **49**, 35–39 (2013). https://doi.org/10.3103/S0003701X1301009X

74. J.D. Major, R.E. Treharne, L.J. Phillips, K. Durose, A low-cost non-toxic post-growth activation step for CdTe solar cells. Nature **511**, 334–337 (2014). https://doi.org/10.1038/nature13435
75. T.L. Chu, S.S. Chu, C. Ferekides, J. Britt, C.Q. Wu, Thin-film junctions of cadmium telluride by metalorganic chemical vapor deposition. J. Appl. Phys. **71**, 3870–3876 (1992). https://doi.org/10.1063/1.350852
76. T. Ferid, M. Saji, Transport properties in gallium doped CdTe MOVPE layers. J. Cryst. Growth **172**, 83–88 (1997). https://doi.org/10.1016/S0022-0248(96)00740-3
77. I.M. Dharmadasa, A.A. Ojo, Unravelling complex nature of CdS/CdTe based thin-film solar cells. J. Mater. Sci. Mater. Electron. **28**, 16598–16617 (2017). https://doi.org/10.1007/s10854-017-7615-x

Chapter 8
Conclusions, Challenges Encountered and Future Work

8.1 Conclusions

The work presented in this book puts together the chemistry, physics, material science, device physics and engineering involved in electroplated semiconductor material deposition and photovoltaic device fabrication, assessment and development. The observation and results are systematically reported in Chaps. 4–7 of this book. The main semiconductor materials explored and presented in this book include ZnS (which is the main buffer layer material) in Chap. 4, CdS (which is the main window layer utilised) in Chap. 5 and CdTe (which is the main absorber layer utilised) in Chap. 6. Also included in Chap. 6 is the synthesis of doped CdTe layers (CdTe/F, CdTe/Cl, CdTe/I and CdTe/Ga). The electronic parameters of fabricated photovoltaic devices as reported in Chaps. 4–6 are reported in Chap. 6.

The ZnS and CdS layer growth and characterisation documented in this book have been published in the literature (see [1, 2]), the CdTe work has also been published (see [3–8]), while the work based on device fabrication was published in [9–15].

The following summations can be made based on the results presented in this book:

1. ZnS layers were electroplated successfully from electrolytic bath containing zinc sulphate monohydrate ($ZnSO_4 \cdot H_2O$) and ammonium thiosulphate (($NH_4)_2S_2O_3$) as precursors of Zn and S, respectively. The crystalline ZnS layers with a thickness of 50 nm were utilised as buffer layers incorporated in the ZnS/CdS/CdTe heterojunction layers with CdS (window layer) reduction.
2. CdS layers were successfully electroplated using thiourea (NH_2CSNH_2) precursor in which there was no S precipitation during CdS growth. The optimum thickness of electroplated CdS window layers incorporated in the CdS/CdTe heterojunction layers is between 100 and 150 nm.
3. CdTe layers were successfully electroplated using both nitrate- and chloride-based precursors which are different from the sulphate-based norm.

© Springer International Publishing AG, part of Springer Nature 2019
A. A. Ojo et al., *Next Generation Multilayer Graded Bandgap Solar Cells*,
https://doi.org/10.1007/978-3-319-96667-0_8

4. Doping of CdTe with F, Cl, I and Ga was achieved with optimal electrolytic bath doping of 20 ppm, 1.5 M CdCl$_2$ (base precursor), 5 ppm and 20 ppm, respectively, based on optoelectronic material properties. Higher efficiencies can be achieved when Cd-rich n-CdTe is used instead of Te-rich p-CdTe.

5. Glass/FTO/n-CdS/n-CdTe/Au (n-n + large Schottky barrier) shows better prospect as compared to glass/FTO/n-CdS/p-CdTe/Au (n-p) junctions of the same CdS/CdTe heterojunction. This was iterated with the effect of growth voltage of CdTe on the fabricated device conversion efficiency.

6. The basic multilayer graded configurations explored in this work include glass/FTO/n-ZnS/n-CdS/n-CdTe/Au and glass/FTO/n-CdS/n-CdTe/p-CdTe/Au with the highest efficiency being 14.1% and 15.3%, respectively. It should be noted that the highest efficiency of 18.4% was observed for glass/FTO/n-CdS/n-CdTe/p-CdTe/Cu-Au devices, but due to instability and reproducibility issues, the results have not been published yet until the electronic parameters of the fabricated cells can be stabilised and further experimentations are performed.

7. The effect of the p-layer thickness of glass/FTO/n-CdS/n-CdTe/p-CdTe/Au was also explored. It was observed that p-layer thickness within the range of ~35 nm gives good rectifying behaviours under dark and better solar cells with higher efficiency.

8. The effects of the inclusion of GaCl$_3$ in the usual CdCl$_2$ treatments were explored and compared to the normal CdCl$_2$ treatment relative to the material and optoelectronic properties of CdTe thin films and fabricated devices. Comparatively, better parameters were observed in the optoelectronic properties of GaCl$_3$ + CdCl$_2$-treated CdTe layers and associated devices. This is the pH value used for growing CdTe layers.

9. The effects of post-growth treatment pH of the normal CdCl$_2$ and GaCl$_3$ + CdCl$_2$ were also explored with preferred optoelectronic parameters observed at pH2 on both counts.

10. The incorporated optimal F-, Cl-, I- and Ga-doping concentration in CdTe shows interesting results with Ga-doping showing more promising results producing material layers with higher electrical conductivity.

Based on the iterated points, the glass/FTO/n-CdS/n-CdTe/p-CdTe/Au configuration incorporating ~120 nm n-CdS, ~1200 nm n-CdTe and 35 nm p-CdTe and post-growth treated with GaCl$_3$ + CdCl$_2$ shows better electronic properties (V_{oc} = 730 mV, J_{sc} = 33.8 mA cm^2, FF = 0.62 and conversion efficiency of 15.3%) and stability. Therefore, the glass/FTO/n-CdS/n-CdTe/p-CdTe/Au configuration is deemed more promising for continued exploration and optimisation.

8.2 Challenges Encountered in the Course of This Research

The major challenge encountered in this work was achieving high-efficiency photovoltaic devices and reproducibility. This can be attributed to several factors including:

1. The control of the electrodeposition process due to the alteration of current density as an increase in the deposition layer thickness increases its resistance.
2. Control and regulation of ions within the electrolytic bath – as a result of depletion in the ionic concentration and the inability to gauge/measure ionic concentration in the electrolyte during layer deposition, thereby reducing reproducibility tendencies.
3. Non-uniformity of electrodeposited semiconductor layers – this is due to the nucleation mechanism of electroplated materials.

8.3 Suggestions for Future Work

To further improve the achieved conversion efficiency, the following itemised suggestions are proposed:

1. Control of concentration of ions in the electrolytic bath using a composition analyser and automated feed pump for replenishing utilised ions in the deposition electrolytes.
2. Explore other graded bandgap solar cell configurations by incorporating suitable wide bandgap p-type window materials as shown in Fig. 2.13a.
3. Explore further alternatives to the reduction of resistivity in both the CdS and CdTe-base materials.
4. Reduction of series resistances of fabricated devices.

References

1. A.A. Ojo, I.M. Dharmadasa, Investigation of electronic quality of electrodeposited cadmium sulphide layers from thiourea precursor for use in large area electronics. Mater. Chem. Phys. **180**, 14–28 (2016). https://doi.org/10.1016/j.matchemphys.2016.05.006
2. H.I. Salim, O.I. Olusola, A.A. Ojo, K.A. Urasov, M.B. Dergacheva, I.M. Dharmadasa, Electrodeposition and characterisation of CdS thin films using thiourea precursor for application in solar cells. J. Mater. Sci. Mater. Electron. **27**, 6786–6799 (2016). https://doi.org/10.1007/s10854-016-4629-8
3. A.A. Ojo, I.M. Dharmadasa, Analysis of electrodeposited CdTe thin films grown using cadmium chloride precursor for applications in solar cells. J. Mater. Sci. Mater. Electron. **28**, 14110–14120 (2017). https://doi.org/10.1007/s10854-017-7264-0
4. A.A. Ojo, I.M. Dharmadasa, Effect of gallium doping on the characteristic properties of polycrystalline cadmium telluride thin film. J. Electron. Mater. **46**, 5127–5135 (2017). https://doi.org/10.1007/s11664-017-5519-4

5. A.A. Ojo, I.M. Dharmadasa, The effect of fluorine doping on the characteristic behaviour of CdTe. J. Electron. Mater. **45**, 5728–5738 (2016). https://doi.org/10.1007/s11664-016-4786-9

6. A.A. Ojo, I.M. Dharmadasa, Electrodeposition of fluorine-doped cadmium telluride for application in photovoltaic device fabrication. Mater. Res. Innov. **19**, 470–476 (2015). https://doi.org/10.1080/14328917.2015.1109215

7. A.A. Ojo, I.M. Dharmadasa, Effect of in-situ fluorine doping on electroplated cadmium telluride thin films for photovoltaic device application, in *31st European Photovoltaic Solar Energy Conference and Exhibition*, 2015, pp. 1249–1255. https://doi.org/10.4229/EUPVSEC20152015-3DV.1.40.

8. I.M. Dharmadasa, O.K. Echendu, F. Fauzi, N.A. Abdul-Manaf, O.I. Olusola, H.I. Salim, M.L. Madugu, A.A. Ojo, Improvement of composition of CdTe thin films during heat treatment in the presence of CdCl2. J. Mater. Sci. Mater. Electron. **28**, 2343–2352 (2017). https://doi.org/10.1007/s10854-016-5802-9

9. A.A. Ojo, I.O. Olusola, I.M. Dharmadasa, Effect of the inclusion of gallium in normal cadmium chloride treatment on electrical properties of CdS/CdTe solar cell. Mater. Chem. Phys. **196**, 229–236 (2017). https://doi.org/10.1016/j.matchemphys.2017.04.053

10. A.A. Ojo, I.M. Dharmadasa, Optimisation of pH of cadmium chloride post-growth-treatment in processing CdS/CdTe based thin film solar cells. J. Mater. Sci. Mater. Electron. **28**, 7231–7242 (2017). https://doi.org/10.1007/s10854-017-6404-x

11. A.A. Ojo, I.M. Dharmadasa, Progress in development of graded bandgap thin film solar cells with electroplated materials. J. Mater. Sci. Mater. Electron. **28**, 6359–6365 (2017). https://doi.org/10.1007/s10854-017-6366-z

12. A.A. Ojo, H.I. Salim, O.I. Olusola, M.L. Madugu, I.M. Dharmadasa, Effect of thickness: a case study of electrodeposited CdS in CdS/CdTe based photovoltaic devices. J. Mater. Sci. Mater. Electron. **28**, 3254–3263 (2017). https://doi.org/10.1007/s10854-016-5916-0

13. A.A. Ojo, I.M. Dharmadasa, 15.3% efficient graded bandgap solar cells fabricated using electroplated CdS and CdTe thin films. Sol. Energy **136**, 10–14 (2016). https://doi.org/10.1016/j.solener.2016.06.067

14. O.I. Olusola, M.L. Madugu, A.A. Ojo, I.M. Dharmadasa, Investigating the effect of GaCl3 incorporation into the usual CdCl2 treatment on CdTe-based solar cell device structures. Curr. Appl. Phys. **17**, 279–289 (2017). https://doi.org/10.1016/j.cap.2016.11.027

15. A.A. Ojo, I.M. Dharmadasa, Analysis of the electronic properties of all-electroplated ZnS, CdS and CdTe graded bandgap photovoltaic device configuration. Sol. Energy **158**, 721–727 (2017). https://doi.org/10.1016/j.solener.2017.10.042

Index

© Springer International Publishing AG, part of Springer Nature 2019
A. A. Ojo et al., *Next Generation Multilayer Graded Bandgap Solar Cells*,
https://doi.org/10.1007/978-3-319-96667-0

Printed in the United States
By Bookmasters